S. HRG. 109–534

SAGO MINE DISASTER AND AN OVERVIEW OF MINE SAFETY

HEARING

BEFORE A

SUBCOMMITTEE OF THE
COMMITTEE ON APPROPRIATIONS
UNITED STATES SENATE

ONE HUNDRED NINTH CONGRESS

SECOND SESSION

SPECIAL HEARING
JANUARY 23, 2006—WASHINGTON, DC

Printed for the use of the Committee on Appropriations

Available via the World Wide Web: http://www.gpoaccess.gov/congress/index.html

---

U.S. GOVERNMENT PRINTING OFFICE

27–461 PDF   WASHINGTON : 2006

CONTENTS

SAGO MINE DISASTER AND AN OVERVIEW OF MINE SAFETY

MONDAY, JANUARY 23, 2006

U.S. Senate,
Subcommittee on Labor, Health and Human
Services, Education, and Related Agencies,
Committee on Appropriations,
Washington, DC.

The subcommittee met at 11 a.m., in room SH–216, Hart Senate Office Building, Senator Arlen Specter (chairman) presiding.

Present: Senators Specter, DeWine, Harkin, and Byrd.

OPENING STATEMENT OF SENATOR ARLEN SPECTER

Senator SPECTER. Good morning, ladies and gentlemen. The Appropriations Subcommittee on Labor, Health and Human Services and Education will now proceed with this hearing, which will inquire into mine safety on coal mines.

This hearing is prompted by the disaster at the Sago Mine on January 2. There have since been two more fatal accidents in West Virginia. I thank the distinguished ranking member of the full committee former president pro tempore Senator Byrd for his requesting the hearing. This subcommittee has been active on the investigation of mine disasters in the past, notably the Quecreek, which occurred in Pennsylvania in the year 2002, and there are very, very important questions on the issue on what has happened at these mine incidents, focusing on Sago.

The Pittsburgh Post Gazette set the stage for our consideration here in a very poignant statement as follows: "The rest of the world will move on—in their reference to the Sago Mine Accident. In the weeks and months to come, there will be other disasters, other wars, other political scandals. But for the families of the 12 men who died in the mine at Tallmansville, West Virginia, for the one who survived and their relatives and friends, for the investigator searching for the cause of the mine explosion, for the people of these coal-rich hills 100 miles south of Pittsburgh, Sago will be a daily litany. Some questions about the January 2 accident may never be answered, but there is understanding to be gained by reconstructing what happened".

We all know that mining is a very dangerous business. As I noted a moment ago, two underground coal mine incidents this January after the Sago disaster. On January 10, there was a miner killed in Kentucky after a roof cave-in. On January 19, two miners became trapped in the Aracoma Mine in Melville, West Virginia. Both of them were fatalities. Last year, the safest year on record, there were 22 fatalities in underground coal mines in 20 separate accidents. Four deaths in my home State, Pennsylvania, three in West Virginia, eight in Kentucky and seven in other States.

The Sago Mine had some 208 citations, orders and safeguards issued against it last year, 2005, with nearly half of these violations categorized as "significant and substantial". The budget for mine and health safety has increased by some 42 percent over the past 10 years, but these increases have barely kept up with inflationary costs. This has forced the agency to reduce staffing, we're told, by some 183 positions over that period of time. Coal mine staffing has declined by 217 enforcement personel in the last 5 years. One of the grave difficulties we have had in maintaining funding for mine safety is the fact that the Department of Labor, which this committee has jurisdiction over for appropriations purposes, is grouped with the Department of Education and the Department of Health and Human Services, and there are many demands on this subcommittee. Senator Harkin, the distinguished ranking member, Senator Byrd, and I, as well as others on this subcommittee have advocated a more realistic budget through this subcommittee, which has jurisdiction over worker safety, jurisdiction over education and jurisdiction over healthcare. We were underfunded in the last budget resolution, and then we sustained a 1 percent across-the-board cut, and I've already put the leadership on notice in the Senate that I'm not going to support a budget resolution that does not adequately fund this department next year or not adequately fund this subcommittee next year. Of course, I hadn't anticipated the Sago Mine Disaster, but we saw from Quecreek, and we've seen from other disasters the importance of having adequate funding. Now, let me yield at this time, it's always a difficult matter whether to recognize the ranking member or the senior member, but I've already had a note here from the generous ranking member, Senator Harkin, asking or instructing that I recognize Senator Byrd first. Senator Byrd, will you accept that recognition?

STATEMENT OF SENATOR ROBERT C. BYRD

Senator BYRD. Mr. Chairman, I do. First, let me compliment you, and Senator Harkin. Let me thank you and Senator Harkin for the great work you do on this committee, and the great work that you have done in the interest of safety and education and other matters that come under the jurisdiction of this committee. You have been very, very great and fine, generous, considerate in this particular situation. Each of us has coal mines in his or her State, and it is very important that the people out there understand that this subcommittee is going to do everything it can by way of funding and oversight to see that the safety and health of coal miners are not underfunded as they have been in the past. I'll get more into that. Shall I do that right now?

Senator SPECTER. Please do, Senator Byrd.

Senator BYRD. I thank you, Mr. Chairman. Fourteen men have died in the coal fields of West Virginia, 14 men in the span of 3 weeks. These deaths, I believe, were entirely preventable, and we owe the families of these deceased and noble, and great, and brave men a hard look at what happened and why. I'm striving to be measured in my statements. I have deep feelings in this matter, but I'm going to try to be measured and careful in my questions, but why was it that the Federal agency charged with safety—the Federal agency charged with safety of this country's miners, failed to protect them. Why was it? Well, let me be clear.

I have no complaint with the brave rescuers who risked their lives to try to save trapped miners, or with the rank and file former miners who work for MSHA in the field. I wish every American could know the total, the absolute total commitment of these folks of these people and understand the dangers faced by rescue teams, whether from secondary explosions like this one at the Sago Mine in Upshur County, West Virginia or the extreme heat of the raging underground fire at the Alma Mine in Logan County, West Virginia. I'm talking today. Let me be clear. I'm talking today about the leadership in MSHA's Washington office have decisions to cut the mining health and safety enforcement budget and staff, and abandon the critical mining regulations endangered of the safety of these miners of Buckhannon and Melville and Lord knows how many other communities in West Virginia and across the country.

Federal and State investigators must explain what happened at the Sago Mine and at the Alma Mine and whether those tragedies could have been prevented. The Congress, however, has a broader responsibility. We must try to determine what is wrong at MSHA and contemplate how to make sure that the leadership of that Agency does its job. Why did it take 2 hours before MSHA was notified of the Sago Explosion and another 2 hours before MSHA personnel arrived at the scene? Why another 1.5 hours before rescue teams arrived and another 5 hours before the first team entered the Sago Mine? The Mine Act requires that rescue teams be readily available in the event of an emergency, and yet, it took 11 hours, 11 hours before the first team could even begin the rescue effort at Sago, 11 hours in the darkness. Where was the help? Is help coming? Our lights have gone out. It's dark. I can't see.

It took 11 hours before the first team could even begin the rescue effort. Meanwhile, 12 coal miners, 12 at Sago, waited in the dark. The horrors, can one imagine? Twelve miners at Sago waited and waited and waited and waited in the dark 260 feet below the surface armed only with a 1-hour breathing device, 1 hour of precious air, grossly inadequate for miners whom we know may have to barricade themselves and possibly wait many hours, even a day, even 2 days, even more for rescue. The miners could not communicate with the teams involved in the rescue effort underway, and the rescuers could not identify the location of the miners. How sad.

Can you imagine the families of those miners waiting, waiting, waiting? Why, in an age of technological marvels, was the only apparent protection for these 12 souls trapped underground a brattice cloth intended to divert toxic gases from their barricade? Now, a short 2 weeks later, the Nation watched as similar horrors emerged from a second tragedy at the Alma Mine. Delayed notification to MSHA of 2.5 hours, no ability to locate the lost miners and no means for any survivors underground to communicate with rescue teams. Put yourself in that situation, how horrible that is, unbelievable in this age.

The administration's budgets have translated into 4-year cuts in mine health and safety enforcement resources and personnel. Let me say that again. It's worth saying. The administration's budgets have translated into 4-year cuts in mine health and safety enforcement resources and personnel. This past week, or this past weekend, President Bush called my house to talk with me on another matter and in passing noted of the Alma Mine fire and asked if he could do anything. Here was the President, President Bush, asking me if he, the President, could do anything. I said, Yes. Stop cutting the coal health and safety enforcement budget. We need more enforcement, not less. We should be demanding a rapid response system in case of a mine emergency. We should be demanding new communications technology and emergency breathing equipment. We must look at the safety and health regulations that MSHA has on the books as well as those that have been discontinued. Those rules govern everything from mine rescue teams to the use of belt entries for ventilation. Let me say that again. Everything from mine rescue teams those rules govern to the use of belt entries for ventilation to inspection procedures, to emergency breathing equipment, to escape routes and rescues and refuges, anyone of which could figure into the disaster at the Sago and the Alma mines. How about that?

What should be done about the pattern of violations that existed at the Sago Mine, and how has MSHA employed its statutory authority to pursue habitual violators? Fines may have been levied, but the sheer number of violations suggests that those penalties are failing to induce mine operators to faithfully comply with the law. Mine safety demands aggressive leadership.

The coal miners in my State, Senator Rockefeller and I speak for them. The coal miners of my State have been doing this kind of work for a long, long time. I should say here again that Senator Rockefeller would have been here today at this hearing, but he could not be because of personal matters, important matters. Senator Rockefeller's heart is right at the center, and he works and hopes and feels like a Senator should whose coal miners are losing their lives. He has worked hard. Mine safety, as I say, demands aggressive leadership. The miners in my State are incredibly brave. Coal miners are a different breed. I know. They will go into mines after a disaster like this and risk their own lives to try to help their buddies, try to find their comrades, their companions. They are brave—brave, and so are their families. They know there's danger there. All the time, they know it. Mine safety demands, as I say, aggressive leadership, and these people are very resourceful when confronted with an emergency. MSHA and the Labor Department are filled with career professionals with hundreds of years of accumulated experience in coal safety.

Given the resources of this Government and the statutory authority of The Mine Act, most mining fatalities should be preventable. We've come a long way, but we have more to do. The Congress must not tolerate complacency about the health and safety of the brave men and women who risk their lives to provide energy, power to run these lights, which make it possible for us here, the energy and power in time of peace and in time of war. Profits should never come before protection for the coal miners.

Did you hear me out there? Profits should never come before the safety and welfare and wellbeing of the brave men and women who go into the bowels of the earth and scratch the coal out, sometimes with their hands, with water seeping from overhead and with danger all around. Politics must never play a role in the enforcement of safety and health regulations. We must not allow the call for action in the wake of this tragedy to recede without stiffening mine enforcement and emergency preparedness.

Senator Specter and Senator Harkin, I thank you again for scheduling this hearing. I hope that this is the beginning of an effort that will yield a worthy legacy for the Sago and Alma miners and for other miners throughout West Virginia and throughout the country. Thank you very much.

Senator SPECTER. Thank you very much, Senator Byrd. As Senator Byrd has noted, Senator Rockefeller could not be here today, he is out of town. Similarly, Senator Santorum could not be here today because he is out of town, and I know both of those Senators and doubtless others would have wanted to have been here.

I might say, by one word of explanation, that the Senate is not in session today although this room has been fully occupied by the Alito hearing for many, many days, so the Judiciary Committee members are here in any event as is Senator DeWine. I want to yield now to my distinguished ranking member. We have a custom of 5-minute opening statements, but the clock got stuck when Senator Byrd was speaking. We'll give him a little more latitude.

Senator BYRD. Thank you, Mr. Chairman.

Senator SPECTER. Senator Byrd has been in the Congress of the United States since Harry Truman was President, elected to the House in 1952 and the Senate in 1958, and it is quite a record, so we have a little more latitude.

Senator BYRD. You have been very kind, Mr. Chairman, very courteous as is characteristic of you. Thank you.

Senator SPECTER. You deserve it, Senator Byrd. Senator Harkin.

STATEMENT OF SENATOR TOM HARKIN

Senator HARKIN. Thank you, Mr. Chairman, and thank you for hosting this hearing. I especially want to thank Senator Byrd for requesting this hearing. Over the years, I've listened to Senator Byrd speak eloquently about the mining communities of West Virginia. We heard it again here this morning. I know that there is no greater advocate for the families of miners, no more dedicated public servant for the people of West Virginia than Senator Robert C. Byrd, so I thank him for requesting this hearing, but also for the dedication that he has exhibited over all his lifetime in fighting for miners not only in West Virginia, but all over the country.

I know that he will continue to hold the folks of Congress on this important matter, and for that we should all be grateful. Although I'm far from the first, I offer my sincere condolences to all the families of the fallen miners, the 12, the 4 and the 2 on Saturday in the Logan County Fire. As the son of a coal miner myself, your families remain in my thoughts and prayers.

My father was born in 1886, started mining coal somewhere around 1910, in Iowa. A lot of people don't know that there were coal mines in Iowa. There used to be a coal mine capital. There's still a lot of coal there, but that's soft coal. They don't mine it any longer. A great coal mine leader, John L. Lewis, many people think he came from Pennsylvania or West Virginia. He actually came from Iowa.

That is where he was born and raised, and that's where he mined coal, and that's where his brother mined coal all his life in Iowa. So we had a lot of coal mines in those days. My father mined coal for 23 years in a coal mine in Iowa. And a lot of coal mines in Iowa. It wasn't till the depression hit that he finally left the mines, but the mines never left him because later in life, he had what we all called miner's lung. We never called it black lung, never heard of that word before. Black miner's lung, every miner had it.

Of course, it didn't help that he smoked all the time too, but he had miner's lung, and that is what finally got him. I can remember as a kid, listening to my father tell me stories of working in the coal mines. Now, you're talking about the teens, World War I and the 20s and going down in these shaft mines a couple of hundred feet and going out in the laterals and telling me about how dark it was and the cave-ins and the miners that would get killed. Trying to get air down, they would pump air down in these mines. He would describe it to me. They would take these rickety old elevators down first thing in the morning before the sun would come up. Sometimes he would come back out of the mines after the sun went down. He never saw the sun during the day. Twelve hours a day they worked down in those mines and never see the sun.

Well, he was one of the lucky ones, but he told me, and I used to see his friends as I was a kid growing up, some of the old miners would come around, people from the old country, the Italians, the Slovenes, the Welsh, the Irish, would all come around, all the old miners, and they would tell their stories. As a kid, I used to listen to these. They were frightening stories. They all would tell the stories of their friends who had died in the mines, and to me, it just painted a very awesome portrait of what it meant to be a miner and how brave these people were in what they did. Of course, that was before any safety regulations or anything.

So, I come at this with that kind of a background, and I wonder why it is—why it is—why it is over—after over a hundred and some years of mining coal in this country, why do we still have mining tragedies like this? Now, I know it's a dangerous occupation. People make mistakes. It's a very dangerous occupation, but still, accidents like these are avoidable. Why—why don't we have, after all this time, better coordinated risk procedures?

Now we have regulations for rescue procedures, but it seems like they more often fall into the breach rather than the letter. Why, after 208 violations in 2005, 95 of them, 40 percent were for serious and substantial violations at Sago, why wasn't something done? Technology. Why? In 1969, the Bureau of Mines, it's not called that anymore, promoted a handheld system so miners could find their exact location, and they could be found. That was used in Utah a couple of years ago in a disaster.

I have a nephew of mine, my sister's boy, he's a miner, goes down everyday in the mines, not coal mines, Trona Mine. Soda Ash, Wyoming, deep mines, they go down a long way. He's been mining there for over 20 some years in that Trona Mine. He's on a rescue team. They have competitions to see how they would set up. They don't have those kind of accidents. Why, in a Soda Ash Mine in Wyoming, don't they have the kind of accidents that we're talking about here? Why is it just coal miners? Why is it just coal miners? Are they less valuable or something? Is there something less valuable about coal?

As Senator Byrd eloquently said, that's what makes our lights light. It makes my timer go down and says I'm overdue. It seems to me that we do have other mines, and they operate more safely, and they have better procedures. Why haven't we done this? It seems to me, and I'll just end on this, Mr. Chairman, it just seems to me that we have ignored this and ignored it and ignored it and put it off and put it off. We've cut the budgets. We don't have the safety procedures. We don't have the handheld devices so that we can find a miner and his exact location. That is possible. It can be done. It's not that expensive. So why haven't we done it? These are the questions that MSHA needs to answer. Thank you. Thank you, Mr. Chairman.

Senator SPECTER. Thank you very much, Senator Harkin. Senator DeWine.

STATEMENT OF SENATOR MIKE DE WINE

Senator DEWINE. Thank you, Mr. Chairman. Mr. Chairman, first, thank you very much for holding this hearing. Our thoughts and prayers go out to the families of the miners who have lost their lives, and we think about them this morning.

Miners are a courageous group of people. Mining is horribly dangerous work. It takes a strong individual to endure the physical as well as the mental toughness needed to work in this industry. Everyday these miners put their lives on the line so that we can have electricity, so we enter a room and flip on a switch, the light does, in fact, come on. Coal plays such a significant role as an energy supply for our Nation. Coal generates more than half of the Nation's electricity. Eighty-seven percent of my home State of Ohio is electric power, is generated by coal. With a 250 to 300 year supply, coal is the only fossil fuel this country has in great abundance. Coal is part of this country's future.

Mr. Chairman, and members of the committee, we simply have to figure out how we can mine more safely than we do today. Obviously, progress has been made, but we have to do better, and these tragedies that this country has suffered, these families have suffered, brings that home to us.

Mr. Chairman, I know our witnesses this morning who we're going to hear in just a moment will help give us insight into what we can do to see this vision move forward, and so I thank you for holding this hearing. And also as a member of the full Health Committee, I'll be working with Chairman Enzi when our committee holds its future hearings as well. So again, Mr. Chairman, I thank you for holding this very important hearing today. I thank our witnesses. And again, my thoughts and prayers remain with the families of the miners that have been lost in the recent tragedies.

Senator SPECTER. Thank you very much, Senator DeWine. We now move to our first panel. We have with us the Acting Assistant

Secretary of Labor for Mine Safety and Health, Mr. David Dye, U.S. Department of Labor, bachelor's degree from University of Texas at Austin and law degree from Franklin Pierce Law Center in New Hampshire. Mr. Dye's accompanied by Mr. Ray McKinney, Administrator of the Coal Mine Safety and Health unit, Mr. Bob Friend, Deputy Assistant Secretary for Mine Safety and Health, no relation to the former Pittsburgh pitcher, and Mr. Ed Clair, Associate Solicitor for Mine Safety and Health.

We've allocated 8 minutes to Mr. Dye for a presentation, which will be joined in by Mr. McKinney. Our standard practice is to allocate 5 minutes for each witness. We have quite an extensive hearing today with good representation on the panel, especially for a day when the Senate is not in session. Thank you for joining us, Mr. Dye, and we look forward to your testimony.

STATEMENT OF HON. DAVID G. DYE, ACTING ASSISTANT SECRETARY, MINE SAFETY AND HEALTH ADMINISTRATION, DEPARTMENT OF LABOR

ACCOMPANIED BY:

>   **RAY McKINNEY, ADMINISTRATOR FOR COAL MINE SAFETY AND HEALTH, MINE SAFETY AND HEALTH ADMINISTRATION, DEPARTMENT OF LABOR**

>   **ROBERT M. FRIEND, ADMINISTRATOR FOR METAL AND NONMETAL MINE SAFETY AND HEALTH, MINE SAFETY AND HEALTH ADMINISTRATION, DEPARTMENT OF LABOR**

>   **ED CLAIR, ASSOCIATE SOLICITOR, DEPARTMENT OF LABOR**

Mr. DYE. Thank you, Mr. Chairman, Senator Byrd, members of the committee. Thank you for this opportunity to appear before you to discuss the work of the Mine Safety and Health Administration and its performance related to the January 2 Sago Mine Explosion in Tallmansville, West Virginia.

My heart is especially heavy today as I appear before you just days after the Aracoma Mine accident in Melville, West Virginia, that claimed two lives. I would like to commend the 21 rescue teams involved in the Aracoma Mine rescue efforts this weekend and the 13 mine rescue teams involved at Sago Mine who put their own lives on the line in the quest to save their fellow miners. Most of these brave rescuers are volunteers. All of them willingly face danger and give generously of their time and in constant training. We cannot thank them enough for their help and their sacrifice; they are the embodiment of American spirit, selflessness and courage.

PREPARED STATEMENT

All of us at MSHA deeply grieve the loss of these miners' lives, and I want to assure the families and friends that we are conducting uncompromising investigations of these accidents. We will uncover the truth why both of those tragedies happened and how we can better protect miners and prevent tragedies such as these in the future. Mr. Chairman, my colleague, Ray McKinney, will now give a short presentation on the events following the explosion at Sago Mine.

[The statement follows:]

Prepared Statement of Hon. David G. Dye

Mr. Chairman: Thank you for this opportunity to appear before the Subcommittee to discuss the work of the Mine Safety and Health Administration (MSHA) and its performance at the January 2 Sago Mine explosion in Tallmansville, Upsur County, West Virginia. We have only preliminary information on the mine accident in Logan County, West Virginia, but I will be happy to share what we know thus far with the committee as well.

It is with the heaviest of hearts that every MSHA employee grieves for the miners who died at the Sago mine and the loved ones who mourn their passing.

MSHA's reason for being is to ensure that miners return home safe and healthy to loved ones at the end of their shifts. That is our mission and our focus, every day. That is our duty to America's miners and the reason we are conducting the investigation with the greatest care and diligence so we can uncover the full truth of why this tragedy happened and how we can better protect miners in the future.

Although MSHA has significantly stepped-up enforcement in recent years and dramatically reduced injury and fatality rates—mining fatalities dropped 33 percent and total injuries dropped 25 percent from Calendar Year (CY) 2000 to CY 2005— the Sago mine tragedy must cause us to carefully assess MSHA's performance, as the Subcommittee is doing today. From CY 2000 to CY 2005, the number of coal mine fatalities dropped 42 percent and the number of coal mine injuries dropped 22 percent, but that is cold comfort to the families of those who are killed, or to miners who suffer injuries. So we can never let up in improving mine safety, and we will not.

In addition to emphasizing increased enforcement of mine safety laws, MSHA has promoted prevention and worked with the mining community—operators, miners, state and local regulators, trade and labor organizations, manufacturers, suppliers and others—to make America's mines the safest in the world. Collaboration on safety training, technical assistance, and education programs is essential and has been productive in helping miners, supervisors and employers adopt the safest and healthiest mining practices.

These efforts complement our strong enforcement regime and our enforcement results from CY 2000 to CY 2005 are substantial:

—Total Citations and Orders issued by MSHA at all mines increased by 6 percent, from 120,050 to 127,682;

—Total Citations and Orders issued by MSHA at coal mines increased by 18 percent, from 58,304 to 68,818; and

—Total "Significant and Substantial" Citations and Orders issued at coal mines increased by 11 percent, from 23,994 to 26,779.

While enforcement went up, the number of coal mines fell. There were 2,124 coal mines in 2000 and 1,982 in 2005 (through the third quarter).

MSHA enforcement personnel have also significantly increased the issuance of withdrawal orders that force coal mine operators who exhibit an unwarrantable failure to comply with the regulations to shut down areas or operations of the mine affected by the hazard. Unwarrantable failure orders are one of the most severe enforcement actions inspectors can take. And unwarrantable failure orders issued to coal mine operators are up 35 percent over the last five years compared to the previous five year period.

MSHA's increased enforcement has also resulted in increased assessments—penalties imposed on mine operators for violating health and safety standards or the Mine Act. From CY2001 to 2005, the number of high-dollar final assessments imposed on all mine operators was 21 percent higher than during the period of 1996–2000. The total dollar value was up by 16 percent during this same period of time. This increase in high dollar assessments occurred while the total number of mines decreased by 8 percent between 2000 and 2005. There is more enforcement, more assessments and more scrutiny of mine operators than ever before.

Under the Federal Mine Safety and Health Act, whether issued in a citation or an order of withdrawal, each violation is assessed a penalty (statutorily capped at $60,000) which must be assessed in accordance with the law's six statutory criteria. The Administration believes that the statutory cap is too low to deter repeat and egregious violations of the Mine Safety and Health Act, and has urged the Congress to increase the statutory cap from $60,000 to $220,000. This would bring the Mine Act's civil monetary penalties in line with those authorized by the Occupational Safety and Health Act. The Administration again encourages Congress to act on this proposal to aid MSHA in enforcing the Mine Act.

The statute itself provides for a graduated enforcement scheme that provides for increasingly severe sanctions for increasingly serious violations or an operator's unwarrantable failure to comply with a mandatory standard. A chain of increasingly severe enforcement actions causing all persons, except those necessary to eliminate the condition, to be withdrawn from that area until the condition is abated, serves as an incentive for operator compliance and is the statute's most powerful instrument for enforcing mine safety.

In every link of the chain, a withdrawal order is limited by statute to the area of the mine affected by the violation or unsafe condition. In some instances, the area affected could be the entire mine, or the area affected could be so critical or integral to the operation of the mine that its closure effectively shuts down the mine (for example, when the area affected includes the hoist used to move personnel and material in and out of the mine, or a blocked escape way). However, as soon as the operator corrects the cited condition and an MSHA inspector terminates the order, the operator is free to re-enter and resume mining work in that area. At the end of the graduated chain of enforcement, even a relatively minor violation can trigger a withdrawal order. A series of severe withdrawal orders affecting critical areas of the mine can inflict a severe economic penalty on the operator and could even lead to an operator ceasing operations. However, the statute does not allow preemptive permanent closure of a mine that has abated all violations.

MSHA enforcement personnel were vigilant in dealing with Anker Mining, the operator of Sago mine until May of 2005 and International Coal Group (ICG), the successor mine operator. As incidents increased at the mine in 2005, so did the issuance of MSHA citations and increasingly severe withdrawal orders, in accordance with the Mine Act. The enforcement hours spent at the mine in 2005 demonstrate the increased attention the mining operation received: MSHA spent 744 on-site inspection hours in 2005 at the Sago mine, up by 84 percent over the 405 hours spent during calendar year 2004. MSHA inspectors were at the Sago Mine on 93 different days in 2005, and often more than one inspector was at the mine. Inspectors issued 208 citations, orders, and safeguards in 2005, including withdrawal orders that shut down areas of the mine on 18 occasions. In addition, between April 27, 2005 and December 15, 2005, MSHA managers and supervisors met with ICG officials approximately 21 times.

MSHA was ramping up enforcement activity at this mine by issuing these citations and insisting that the cited hazards be promptly corrected. Each one of those citations represented a hazard that was required to be eliminated within a fixed abatement time specified by the inspector—and the inspector followed up to ensure the hazards were indeed fixed. That meant extra inspector time in the mine, and inspector time talking with mine operators, management and supervisors to ensure that those hazards were indeed corrected. Sago Mine management also received substantial engineering consulting services from MSHA's Office of Technical Support on major issues at the mine including water removal, roof control, dust control and other problems. We were beginning to see positive results as the all-injury incidence rate at this mine dropped from 55.8 in the 2nd Quarter to 8.3 in the 3rd Quarter.

While there were many problems at the Sago Mine, liberation of large quantities of methane had not been among them. Methane is explosive between 5–15 percent. At one percent, MSHA requires operators to take action. At the Sago Mine, the highest readings never exceeded two-tenths of one percent, which is one-twenty-fifth of the explosive level. MSHA's routine monitoring of methane in the mine never warranted a citation for excess accumulations of methane gas.

At this time, we have no information that would suggest that the explosion was related to a recurrence of any of the conditions that were required to be abated before the explosion. We will, of course, aggressively pursue these and all other potential causes in our joint accident investigation.

Until the joint investigation team can safely enter the mine to thoroughly examine the site, we will not know the answers to these questions.

It is a sense of duty, fellowship and a strong measure of heroism that mine rescue teams took with them into the Sago Mine in the quest to save their colleagues. Thirteen teams comprised of 109 dedicated team members participated in the rescue and recovery operation, putting their lives on the line for their fellow miners. Their bravery and dedication are emblematic of the 314 mine rescue teams around the country. It is noteworthy that the vast majority of rescue team members are volunteers—good Samaritans in the highest American tradition. It is extraordinarily courageous and generous to take on the responsibility of such perilous rescue missions, knowing that heartache may mark the end of the day. We cannot thank them enough for their help and their sacrifice.

Some observers have raised concern that the Sago explosion occurred at 6:30 a.m. and the first rescue team did not enter the mine until 5:25 p.m. The delay was out of concern for these rescuers' lives. MSHA and the mining community have a history of hard-learned lessons on the peril of rushing into mine accident scenes.

Those lessons were at the cost of the lives of rescuers who went into mines too quickly and died in rescue attempts. As recently as 2001, 12 miners in Alabama attempted to rescue a miner gravely injured after an explosion. Those miners were killed by a secondary explosion. In 1976, 26 miners lost their lives in mine explosions in Overfork, Kentucky. Fifteen miners were killed in the first explosion and 11 miners died in a second explosion, including 3 Federal coal mine inspectors.

At 5:25 p.m. at the Sago Mine, the carbon monoxide and methane readings finally started trending downward to a level where MSHA, West Virginia's state mine regulators and the Robinson Run No. 95 rescue team felt we could take a carefully managed and calculated risk allowing rescue teams to enter the mine.

Mr. Chairman, we at MSHA are determined to find out how this tragedy occurred. While I cannot comment on substantive aspects of the accident investigation, I can promise you that it will be thorough and meticulous. The investigation will examine the circumstances of the accident, the circumstances of the rescue efforts, and every other germane piece of information. All of the data and information collected will be carefully scrutinized and analyzed, and a thorough report prepared and made available to the public. We will finish this investigation as soon as possible, consistent with the need for accuracy, and we will promptly take any action indicated by the investigation's results to improve safety and health in America's mines.

As standard operating procedure, MSHA also conducts an internal review after every major accident. We will carefully examine whether MSHA followed its own policies and procedures with respect to the accident, including the enforcement activities preceding the accident. This report will be provided to the committee and made public on our web site. We view this internal review process as an opportunity to take a hard and honest look at how we do our job and use that information to improve our performance on behalf of America's miners.

Miner health and safety is our bottom line and our only priority. Thank you.

Senator SPECTER. Thank you, Mr. Dye. Mr. McKinney.

STATEMENT OF RAY MCKINNEY

Mr. MCKINNEY. Thank you. This is a very brief presentation, and if you would draw your attention to the power point projector on the right. This is the service area of the Sago Mine that depicts the drift portals and the surface structures. To the right of those portals, you'll see an earthen embankment and that's where the mine offices were located when we set up the command centers. The Sago Mine was about 2.5 miles deep, and the mining height was 5 to 6 feet.

This is an overview of the mine map. The mine has two working sections where miners travel each day to produce coal. Those working sections are the First Left and Second Left. There are extensive shelter areas in the mine. Those are areas that active workings are barricaded and built with proof seals to isolate them from active workers where miners are. Those sealed areas are identified in red in the lower part a little bit farther up, and the most recently sealed area was 3 Left, just in by 2 Left.

The red line indicates the route of travel for the miners to the two active sections each day. On January 2, 2006, the mine was examined by company mine examiners and deemed safe to operate in the interim. Miners went—began to enter the mine at approximately 6 a.m. At approximately 6:30 a.m., an explosion occurred underground. All of the miners exited the mine except for the 12 miners that had traveled toward the 2 Left section. The mine examiner, who remained underground, was traveling from 1 Left to 2 Left. So, we had 13 miners unaccounted for.

An initial rescue attempt by mine management failed when they encountered heavy smoke and excessive concentrations of carbon monoxide in the 2 Left section. MSHA was notified at 8:30 a.m. in the morning by telephone, and we responded, mobilized our force and were on the ground at 10:30 a.m. at the mine. When we got there, the air exiting the mine portals contained in excess of 500 parts per million of carbon monoxide and indications of methane gas. Carbon monoxide is an indicator of a fire or heat. Methane gas, of course, is inherent to mining explosions, and is explosive when it constitutes between 5 to 15 percent of the air in a mine. The carbon monoxide was trending upward, which meant we had indications of a potential fire underground. By noon, the level had reached 2,600 parts per million.

Also at noon, we had to remove people from the mine offices due to the high and elevated levels of carbon monoxide (CO). It was not safe to enter the mine until the CO, or the carbon monoxide level, showed signs of trending downward and stabilizing. We have a rich history of second explosions after an initial explosion. In 2001, we had the Jim Walters event where 12 miners were trying to rescue a fallen miner after a first explosion. All were fatally injured during the second explosion. In 1976, at the Scotia Mine, 15 miners were killed in an initial explosion. During the second explosion, another group was going in, and 11 miners lost their life.

It's a very serious situation when you send people underground after an initial mine explosion. You have to know you're doing the right thing. Air ventilation is generally destroyed after the first explosion. Methane gas has an opportunity to build quickly, and you have indications of fire. The expansion of gases creates hot spots or fires in the coal mine. At 5:25 p.m., the levels had dropped to the point that the teams were sent underground. This was not an ideal situation. This was a calculated risk when we sent those teams underground, and we made contingency plans that if the CO continued to rise again, we would evacuate the mine.

The teams explored the mine for the next 24 hours, encountering water accumulations, destroyed ventilation controls and electrical power sources and circuits. The teams had to methodically explore the mine keeping attention on the air entry return side to make sure what the CO concentrations were leveling out to, and also looking for potential fires. It was important also to look on the intake air entry side in case there was a fallen miner who had tried to exit the mine.

At 4:21 p.m. on January 3, the teams had advanced to the 58th crosscut. That is the area just out by the intersection of 2 Left, the number 4 belt entry. At that point in time, we had not explored the 2 Left section. We'd found the seals blown out in the 3 Left section, which was the one that was most recently sealed. We had not explored that area but we decided, given the time frame, that we needed to extend ourselves on into 2 Left.

The teams began to travel toward the 2 Left faces and found the section mantrap at number 10 block. We also found evidence that indicated the miners had donned their self rescue devices. We asked the teams to advance on into the faces of 2 Left, which is a distance of approximately 2,000 feet, looking for the miners. We rarely ever go more than 1,000 feet with the mine rescue team, we move our fresh air up to those people and allow them to move on. If we have a problem beyond that, it's very difficult to get the team back to the fresh air base, but the teams worked with us in order to advance themselves 2,000 feet.

The miners were located in number 3 entry face area, and the first report came at 11:46 p.m. that all 12 miners were alive. At 12:30 a.m., about 45 minutes later, a second report came out. It was changed that there was 1 miner alive and 11 miners who were deceased. We continued to work underground, and we brought the surviving miner out at 1 a.m. and rendered medical attention to him, and he was transported to the hospital.

We asked the teams to work under oxygen and continue to pick up the other miners and bring them to the surface. It was a long, arduous task through the night. At 9:55 a.m. the next morning, all the miners were extracted from the mine, and that ended the recovery operation.

There's been some question about the communication issues underground. As the honorable Senator said, mine rescue team members are to be honored and their work and their bravery. These teams stretched themselves and pushed the envelope by going in excess of the normal distance in order to try to save life. If you look at this particular slide, you'll see that our communication systems, and these were handheld radios. Normally your sound-powered systems from the mine rescue only go 1,000 feet, so we had to use handheld radios. They work well in the line of sight. They don't work well around 90-degree turns, so we had one set up near the face, one at number 9 crosscut, one at number 58 crosscut and then one at—I'm sorry, one at number 59 crosscut and one at number 58 crosscut. This information has been transferred at least five times, and at least in the first four, it has been done through a mine rescue apparatus face piece with a speaking diaphragm. So I think that it's understandable that those people were doing everything they can, and you create situations for communication when you stretch out that far. Thank you.

Senator SPECTER. Thank you, Mr. McKinney. We'll now start 5-minute rounds for each of the senators on the panel, and I would like everybody to stay within the time limits. We can go to a second round if necessary, and I will stay within the limit myself.

Mr. McKinney starting with you, there was a report that 11 of the miners had been reported alive, 12 miners reported alive, and it was especially cruel and to then find out that it was wrong, that they had died. Was there any way that report, false report, could have been avoided to save so much pain and suffering?

Mr. McKINNEY. I think it is sad, and I think it saddened all of us. There was a lot of people who felt very good for a short period of time and then were deeply saddened by that loss.

Senator SPECTER. Could it have been prevented?

Mr. McKINNEY. As I said earlier, those communications were coming from the underground. They were relayed several times before they reached the surface. We always ask repeat that, and you get a response back what you have, but by the time it reaches the surface, you have to go back through six people again, and I think that's exactly what happened.

Senator SPECTER. Let me move on to another question because of limited time. The Sago Mine had 208 citations in the year 2005. Was any of those citations a causative factor in this disaster? Any of the facts uncovered in the citations a causative factor for this disaster?

Mr. MCKINNEY. I think we're still investigating the disaster itself, and that will be determined by the investigation team if there's anything there we saw through the course of those inspections.

Senator SPECTER. Is it a possibility that one or more of these violations might have been the cause of the disaster?

Mr. MCKINNEY. All of those 208 violations had been abated and corrected. The only outstanding violations that we had, had to do with roof conditions and tunnel liners being cushioned, and things of that nature. The other conditions had been corrected.

Senator SPECTER. I noted that these 208 citations resulted in fines of slightly in excess of $27,000. Was that sufficiently tough punishment for so many citations, Mr. Dye?

Mr. DYE. Nine of the most serious violations have been appealed. Those that haven't been assessed yet could double or triple the amount already assessed.

Senator SPECTER. Was the $27,000 in fines sufficient for those which hadn't been subject to appeal?

Mr. DYE. Those assessments are determined through a formula under our guidelines.

Senator SPECTER. We're going to take a close look at that, and I would like to have your one-by-one itemization of all of these violations specifying exactly what was involved and what penalty was imposed so this subcommittee can make an evaluation.

Mr. DYE. We will supply that for the record.

[Furnished as questions for the record.]

Senator SPECTER. Pardon me for moving on, but there's very little time, and we want to pinpoint the questions and the answers. Your unit had its funding reduced by $2,800,000 by virtue of a 1 percent across-the-board cut. Had the allocation remained without the 1 percent cut that this subcommittee had put into effect, we would have met the budget request for the administration. But with the 1 percent cut, there was that reduction of $2,800,000, and I note that you've lost 183 positions as a result of budget insufficiency. Was that cut causative of the reductions in personnel, and were the reductions in personnel a causative factor in the Sago Mine tragedy?

Mr. DYE. Please allow me to answer the last question first. No, I don't think so. We have dealt with rescissions for several years in a row. Our strategy has been, where we have had to reduce personnel, to try to reduce them in administrative and support positions, to try to reorganize in a way that maximizes our ability to keep our enforcement personnel out there. In fact, the Congress directed us to do that, and we have tried to manage them that way.

Senator SPECTER. We would like a specification of what you would have done with the additional $2,800,000 had you been allocated that, and we would also like a written followup as to what were these 183 positions which you had to cut and what the impact has been on those cuts in terms of what your ability is to improvise mine safety.

In the 16 seconds left, I have a final question for you. I note that you have stated your intention to hold a hearing. Why the delay? These hearings are very important. That is why we responded to Senator Byrd, January 2 incident. We're still in January, and this requirement that the witnesses be interviewed in the presence of company coal officials, if true, why? Isn't there a likelihood that they would be able to be in a position to give more candid testimony if they weren't interviewed in the presence of the company officials?

Mr. DYE. My understanding is under the Mine Act, the company can be present; miner representatives can be present also. It is important to get interviews if you can before you have a public hearing. You do get more candid interviews that way.

Senator SPECTER. Senator Byrd.

Senator BYRD. Thank you again, Mr. Chairman, and thank you, Senator Harkin. Why did it take 2 hours for MSHA to be notified by the Sago explosion?

Mr. DYE. We don't have that answer yet. That's one of the things—you mean by the company, sir? That is one of the things we're going to be investigating.

Senator BYRD. How long do you think it's going to take to find out the answer to that question?

Mr. DYE. The investigation team is in the process of interviewing witnesses. That is a rather thorough process. That should be accomplished in a matter of a few weeks, and then there will be a hearing. As soon as our investigation teams can enter the mine, they'll need to go into the mine and do all their forensic work. Then after that, they'll have to write a report. It's a very careful, thorough process, and we will get that report out and make it public to this committee, to the public generally just as soon as possible, Senator. I promise you that.

Senator BYRD. Well, Mr. Secretary, I think you should know the answer now. That would be the first question, why did it take 2 hours for MSHA to be notified, 2 hours. Then why did it take 6 hours for rescue teams to arrive? Why is a rapid notification and response system not available? Should that mine have been closed with all of those citations? All of the breaking and wearing down, and tearing down and easing the regulations, why?

Mr. DYE. Please allow me to answer the last question first, Senator. The Mine Act has substantial penalties and more importantly, and people don't always recognize this, it has the authority to close areas of the mine that are affected by violations. We did that 18 times until each of those violations were abated. As Ray mentioned, there were a couple of violations that were outstanding at the time, that were in the process of being abated, but The Mine Act does not contain a provision that allows preemptive closure of a mine for accumulated bad acts or a permanent closure.

Senator BYRD. Let me cut right to the chase. Do you think that there's too much cronyism in situations of this kind? Do you think that the regulations and penalties are too soft? Should there be— I wonder if the enforcement procedures are enforced without fear and without favor. Is there a cronyism? Is there a possibility that cronyism develops between the industry and the Federal agencies that has jurisdiction over the safety of these people?

Mr. DYE. Well, there certainly isn't any cronyism between me and anyone. I'm not aware——

Senator BYRD. Will you say that again?

Mr. DYE. There's no cronyism between me and anyone in the industry. With respect to the agency itself, particularly our coal mine division, I would note that our inspectorate have to have 5 years' experience in the industry. They're all miners, sir. They come from mining families. They come from mining communities. They have a fire in their belly to protect miners, and they go out and do that. I encourage them to do that fully.

Senator BYRD. I have no quarrel with that.

Mr. DYE. Yes sir.

Senator BYRD. We know those people are brave and courageous.

Mr. DYE. Yes, they are.

Senator BYRD. But what I'm asking is, is there too much opportunity here for cronyism? I wonder if the regulations are being enforced without fear or favor, and the fines aren't being levied. Are they heavy enough? They don't seem to be. Too often, industry just seems to pay the fines and go right ahead and keep doing the wrongs. What do we have to do to fix that? That seems, in part, to have been the case here.

Mr. DYE. The administration has sent to the Senate and will to the House, when it reconvenes, a proposel to raise the penalties for flagrant violations to $220,000.

Senator BYRD. Is that just a tap on the wrist?

Mr. DYE. No, I think that's substantial. The other thing I would like to mention is with respect to withdrawal orders and closing down sections of the mine. That can be a very, very powerful tool. In my opinion, even more powerful than citations because if you close down a production area until the hazard is abated, that can cost a company anywhere from $50,000 to $150,000 in lost revenues for a single shift. So we did that 18 times, and that is a way that you can get the attention of an operator very quickly.

Senator BYRD. Mr. Chairman, my time is about up, maybe overdone. Thank you.

Senator SPECTER. Thank you very much, Senator Byrd. Senator Harkin.

Senator HARKIN. Thank you, Mr. Chairman. Mr. Dye, in December 2002, you were the deputy at that time, I believe.

Mr. DYE. No, I wasn't.

Senator HARKIN. You weren't then?

Mr. DYE. No, I was a Deputy Assistant Secretary in the Employment and Training Administration at that time.

Senator HARKIN. When did you come to MSHA then?

Mr. DYE. In May 2004.

Senator HARKIN. In December 2002, MSHA withdrew a proposed regulation meant to encourage mine companies to host mine safety teams on site. It was a proposed regulation, not 2 hours away, but on site. It was withdrawn, I think, by Mr. Lauriski, David Lauriski, with the statement that MSHA "plans to evaluate non-regulatory alternatives". That was 3 years ago. What did you come up with?

Mr. DYE. I would defer to Mr. Clair if you don't mind because he was there at that time. My understanding of that proposal centered around the idea that you would rebate some fines and give that money to the companies to encourage them to have mine rescue teams. I think that after some thought, pretty much everybody thought that that wasn't a good idea because that would undercut the value of the fines for enforcement purposes.

Mr. CLAIR. Senator, I would only add to what Mr. Dye has said that that proposal was something that the agency had under consideration. I do not believe it had been formally proposed. It was a work in progress, and the theory behind it had substantial problems to it. It looked to economic incentives to mine operators that essentially were funded out of civil penalties, and that is a serious legal problem. There were also difficulties when you streamline an existing standard like mine rescue standard that is quite complex and comprehensive. If you streamline that, there's a potential for reducing protections that already exist.

Senator HARKIN. There was also an advance notice of proposed rule making in 1997 that was issued to improve the oxygen devices worn by miners like the ones at Sago. The self-contained rescue devices, as we know, provide an hour of oxygen, they were required in 1981, 12 years after they were invented, that every miner must be equipped with one. This regulation was dropped from the agenda in December 2001. The regulation would have addressed the manufactured quality of the devices, the training of miners on its use. There's one other part of that regulation, that my staff showed me that was very interesting.

The regulation, the proposed regulation, would have addressed the need for updated technology including personal emergency devices, underground text messages devices. Why was that dropped in 2001?

Mr. DYE. Well, getting back to 1997 is really getting backed——

Senator HARKIN. It was dropped in 2001.

Mr. DYE. Yes.

Senator HARKIN. December 2001.

Mr. DYE. Yes.

Senator HARKIN. Why was it dropped? Do you even know? If you don't know——

Mr. DYE. Well, with respect to the self-contained rescuers, there was a question about the service life of those devices, and I'm told that it simply was not evidence that the manufacturer service life should be reduced.

Senator HARKIN. Say that again.

Mr. DYE. It concerned the recommended service life of those devices. In other words, how long you should keep them or could keep them for—before you use them.

Senator HARKIN. The reason I brought up—brought it up, the other part of it, in 1998, there was a mine fire at the Willow Creek Mine in Helper, Utah. Personal emergency devices—which is in that proposed regulations, instantly and wirelessly received emergency message broadcast to each miner on the screen. Forty-five miners used that to escape the fire. Seven years later, only 17 mines in the United States utilized this technology. Why?

Mr. DYE. Well, I talked to my tech support folks about that, and there are, as you said, some mines that use it. Some of them, candidly, are enthusiastic about it. Some of them have had a number of problems, including reliability issues. That is, I believe, a one-way communication device. We're looking to see if that can be made into a two-way communications device, which would be much more useful.

Senator HARKIN. Well I just wanted to read here, here's 12/01. MSHA's withdrawing this entry from the agenda in light of resource constraints, and changing safety and health regulatory priorities. Resource constraints, you know what that means, we know what that means.

Senator BYRD. That's money.

Senator HARKIN. I don't know what changing health and safety regulatory priorities mean, priorities. Well it seems to me again I harp on this, because there is technology out there that can be used. My time has run out too, but I want to get back to that Mr. McKinney, you talked about not getting around the corner. There's easy simple technology to take care of that, very simple easy technology. I realize what they had couldn't get around the corners, it's line of sight. But, that can be changed in a matter of minutes with technology.

Thank you, Mr. Chairman.

Senator SPECTER. Senator Harkin, did you say you wanted to get back to that, if you want a few more minutes?

Senator HARKIN. No, that's okay.

Senator SPECTER. Senator DeWine.

Senator DEWINE. Mr. Dye, let me kind of follow-up on what Senator Harkin was asking you, and what some of the other of my colleagues have been asking you, I'm not sure I quite understand your procedure that you're going to follow in order to conduct the investigation, and then come forward with recommendations. Where are you going to go from here? What is the time line, how long is the investigation do you think it's going to take? So, you can tell us what happened, and then when are you going to come forward with recommendations, and what's the process that you're going through in regard to how you're going to come forward with those recommendations, because that's frankly what we're interested in.

What we're interested in, is where do we go from here? That is what Senator Harkin was getting at. He's talking about the technology that could be utilized. We've all read different experts talking about the technology that could be used. We're kind of interested in what you have to say about it.

Mr. DYE. Well with respect to the investigation, first of all——

Senator DEWINE. Well yes, and then where are you going to go from there? Because that is the bottom line, is making it safer for coal miners. Obviously, that's what we're all interested in. That's what you're interested in.

Mr. DYE. Absolutely sir.

Senator DEWINE. That's what everybody's interested in.

Mr. DYE. Yes sir. We appoint an investigative team in these kinds of matters that is independent. The Lead Investigator designs the investigation; they have to be custom designed because there's a lot of issues. One of the first is, they have to get in and look at the accident scene. They couldn't do that for a long period of time, because they had to clear out a lot of water and gas out of the mine. We still have mine rescue teams.

Senator DEWINE. I only have a couple of minutes. What's the timing on that?

Mr. DYE. At the same time, the investigative team has to interview witnesses. However many they have to do, and they can do about three a day, because they're fairly extensive, and when they've interviewed all of those, and they have that information, they will hold a public hearing, in which they're in the process of planning now. Then they will write a report, and that will all be done as rapidly as possible. But because we don't know the conditions in the mine, and how much time they're going to have to spend in there, it's hard for me to give you hard time line, Senator.

Senator DEWINE. Then what's the process for your recommendations, the administrations recommendations about how we move forward?

Mr. DYE. This is an accident investigation, so the recommendations will be specific to that situation. By the way, we also do an internal investigation to see if we followed our own procedures. Then from that separate inquiry, if we didn't follow our own procedures, or there's ways to optimize them, the internal review team will also write a report. We will adopt those recommendations fairly quickly.

In the case of the Jim Walters accident in 2001, the Agency adopted an emergency regulation very quickly. I want to add with respect generally to mine rescue, we are in the process of publishing, and have put up on our website today, a request for information in the Federal Register.

What we're seeking is information from individuals all over the country, or all over the world for that matter, on mine rescue technologies. These are emerging technologies, new things that are popping up all the time. So, we would like to take a snapshot of what's available now, look at it, evaluate it, and see what is out there.

Senator DEWINE. What is you time line on that? That's good.

Mr. DYE. I believe that is a 60 day comment period going to March, and then our technical folks will have to evaluate all of the comments, and that depends on the volume, and other things.

One thing about these technologies, you can't use off the shelf technologies. Any electrically powered device, including a battery device, has the potential to cause an explosion in a gassy mine. So they'll have to be virtually redesigned, so they won't do that. It is an expensive process, and manufactures have to be willing to do that. I couldn't take my cell phone, or a camera, or an MP3 player, or anything like that into a gassy mine. That would not be permissible, because those things have not been designed so that they would not create a spark.

Senator DEWINE. Mr. Dye, my time is about up. I appreciate your comments. I think what you're hearing from this panel, at least you're hearing from this Senator, is that my constituents expect out of this tragedy something positive to come. And what the positive is, is something specific, and specifics, and the specifics lead to better mine safety, and that's what we look to you to come forward with.

To get back to what Senator Specter was talking about, if we're talking about additional money, we expect you to come forward and tell us that. We want to know how that lines up with the dollars and cents.

Mr. DYE. Yes, I can tell you that there will be additional funding for safety technology—mine safety technology in the President's 2007 budget.

Senator DEWINE. I thank you, my time is up. Thank you.

Senator SPECTER. Thank you, Senator DeWine. Thank you Mr. Dye, Mr. McKinney, Mr. Friend, and Mr. Clair. Mr. Dye and Mr. McKinney, I'm advised that you would like to leave. Your presence will be required here for at least 1 more hour while we move ahead with the next panel. Questions may arise, and we want you here to answer the questions; we've assembled four Senators who may need your responses.

Mr. DYE. Senator, we've still got a mine fire going. We have a rescue team that's in the Sago Mine. We have another mine fire, in which no one was hurt, burning in Colorado. We have really urgent matters that we need to go back to and attend. We have been diverted by dealing with these matters. We were happy to prepare for the hearing, but we really need to get back, and attend to all of this. There are 15,000 mines in the United States, and we've got some really pressing matters.

Senator SPECTER. Well Mr. Dye, I can understand the press of your other business. It may well be that some of the Senators here have other pressing matters too, it may well be. We're looking at the Alito hearings tomorrow in the Judiciary Committee, a matter of enormous importance. Senator DeWine and I have to be there. We're looking at the proceedings under the PATRIOT Act, which expire on February 3, which we have to come to a conclusion on. We're preparing for hearings on the National Security Administration, NSA, about potential excesses in power on the part of the President. We're in the final stages of preparing immigration legislation, which is an 11 million person problem for this country. We're about to go to the floor for the Asbestos Reform Act, and I think Senator Byrd, and Senator Harkin, and Senator DeWine have other matters. So, we don't think we're imposing too much to keep you here for another hour.

If you told me your hearings were going to start tomorrow, or there was some pressing business, we'd say you don't have to be here, but that's the committee's request. You're not under subpoena.

We'll now move on to the next panel. Mr. Cecil Roberts, Mr. Ben Hatfield, Mr. Chris Hamilton, Mr. Davitt McAteer, and Mr. Bruce Watzman. Our first witness on this panel is Mr. Cecil Roberts, the International President of the United Mine Workers of America. Mr. Roberts is a sixth generation coal miner. Both of his grandfathers were killed in mines before he was born. He's a graduate of West Virginia Technical College, where he also received an Honorary Doctorate in Humanities.

Mr. Roberts, you come to this hearing as a distinguished President of the United Mine Workers of America, and going back six generations is very impressive.

Mr. ROBERTS. Thank you sir.

Senator SPECTER. I have a little trouble going back two generations. I only knew one of my four grandparents. So, you have quite a lot of longevity. As announced earlier, the allocation of time is 5 minutes per witness, and we'll begin with you Mr. Roberts.

STATEMENT OF CECIL E. ROBERTS, INTERNATIONAL PRESIDENT, UNITED MINE WORKERS OF AMERICA, FAIRFAX, VIRGINIA

Mr. ROBERTS. Well first of all Senator, thank you for agreeing to have this hearing, and your long years of service, and dedication to coal miners in this country, and there are two other very distinguished Senators I've had the opportunity to work with, and of course the person who lead the fight for coal mine health and safety legislation in 1969. My senior Senator from West Virginia, I've worked with Senator Harkin, and I've had the pleasure of working with Senator DeWine, too.

First, I think it would be appropriate, and I will say it for all of us, that our hearts go out to the families, not only of Sago Mine, but most recently the terrible tragedy that occurred, and ended over the weekend in Logan County, West Virginia.

In my prepared remarks, we never mentioned the situation in Logan, because we did not know at the time that you had requested us to get this information to you, the unfortunate outcome of that tragedy. Let me say that I'm probably going to take a different approach to this as I think that you might expect to represent the coal miners in this country as to what I think needs to be done, and what should've been done, and what is the possibilities for the Congress to do here.

First of all, everyone is now for oxygen being available throughout the coal mine. Well, we've known for many, many years that this was a necessity. I'm not talking about the United Mine Workers when I say we. I would suggest to you that any expert in this industry, if you had asked them yesterday, 1 year ago, 2 years ago, 6 years ago, they would've said to you, if you want to give coal miners the best opportunity for survival, then you provide oxygen in the working places of that coal mine, and you provide it throughout the coal mine.

Now the question is not only should we do this, and require it, and should Congress act, and give a mandate of some type that that be done. I think it is appropriate for the question to be asked, why is it as we come here today, that that has never been done? The question was asked previously here, could these tragedies be prevented? That's one question. There is another question that should be asked and posed here, what is the best opportunity for a coal miner to survive, if and when tragically this occurs again? It's a relatively simple question when it comes to oxygen. If I told everyone we're going to cut off the oxygen to this room right here, and it's going to disappear, there will be no more oxygen, but we're going to take a vote here today, if we want 1 hour of oxygen available to us, or 2 hours, I know what the vote would be.

Second of all, when I started working in the coal mines in 1971, we had a telephone line running to the outside. This was 30 some years ago, to the face. Today if you go in a coal mine, that is what you find in most instances, one telephone line from the surface to the working sections of the mine, now there is some things that could be done, and we've been contacted by the way, by the Pentagon, they don't know if they have the expertise, or the technology to solve this problem for us. But, they took the initiative to phone United Mine Workers, and suggest that we should be talking to them, and we would urge the Senate, this committee, to do that.

There's other things we could do. We could put that portable hand held device that we were talking about the Mine Rescue Team's using. It is only good for 1,000 feet, but for a very minimal cost, I should report to this Committee we could do that. If we could communicate with miners on the wrong side of these disasters, they would know what their best opportunity for survival might happen to be. There's two things we do know, and I don't believe there will be any disagreement amongst anyone on this panel. Had the miners at Sago had more oxygen, and had they had the ability to communicate with the outside, their lives could have possibly been saved. But we do know one thing, it would certainly have given them the best opportunity.

This issue of mine rescue teams, I must admit to this panel today, and I don't want to appear to be too frustrated, and too angry here, but I am. We have known for years that we need additional mine rescue teams in this country, and the problem isn't getting better, it's getting worse. Because, as people my age retire from this industry, who have life long experience as mine rescue team members, then I submit to you, we're losing those experienced mine rescue team members. We've known this for many years, but we have not corrected this problem.

PREPARED STATEMENT

Let me submit one final comment, and I realize that this may be getting into an area that may offend some people, but in 1969 Mr. Chairman, Congress decided this industry was incapable of policing itself, and established MSHA as an independent body to protect the coal miners of the United States of America. In 2001, we put the coal industry in charge of this Agency, and we have submitted 17 rules that were withdrawn, in 2001 or later, that would have protected coal miners in the United States of America, and there are other issues that we could raise with you today, but I realize my time is up, but I did not want to leave here with that on my heart, and on my mind. Thank you.

[The statement follows:]

PREPARED STATEMENT OF CECIL E. ROBERTS

I thank you for this opportunity to appear before Congress. I only regret that I have come to speak on the heels of the terrible tragedy that befell the Sago miners. Our hearts and our prayers go out to the miners' families, their loved ones and their communities. We also wish to extend our deep appreciation for the mine rescue teams that participated in the Sago Mine rescue efforts, and to the Federal and State inspectors and UMWA safety committee members who travel and inspect the mines regularly, working tirelessly to protect the nation's miners.

Today I will be reviewing how current mine safety laws came into being; problems with the enforcement of the current laws and regulations and how those deficiencies have contributed to make coal mining one of the most dangerous industries in the nation; how we have the knowledge and ability to substantially improve miners' health and safety; and what Congress can do to help improve miners' health and safety.

For 116 years, the UMWA has been unwavering in its efforts to enhance miners' health and safety.

However, too often remedial activity follows only after another tragedy has focused our nation's attention: in 1968, 78 deaths at Farmington, West Virginia lead directly and quickly to passage of the Coal Act in 1969; it was then expanded to other mining industries and re-named the Mine Act in 1977. Since the Coal Act was passed, fatalities in coal mining have decreased dramatically: while over 300 miners died in 1968, the year before the Coal Act was enacted, fewer than 100 miners have perished in any single year over the last 20 years, since 1985. While increased mechanization has meant fewer miners are engaged in coal mining, the fatality rate has also dropped significantly. This is commendable; but we can and must do much better. Mining remains the second-most dangerous industry in this country.

This nation possesses the knowledge and ability to substantially improve miners' health and safety, and to reduce the fatality rate. We can direct some of the national attention generated by the Sago tragedy to enhance health and safety conditions for all coal miners today, and the generations to follow.

Unfortunately, what happened at Sago did not really surprise me. Indeed, the underground coal industry has experienced tragedies, as well as near tragedies, on a recurring basis. In just the last few years the underground coal industry experienced these large-scale, well-publicized, events:

Jim Walters Resources #5 mine explosions; 13 fatalities.—There was a terrible series of events on the evening of September 23, 2001, two short weeks after 9/11. What happened there eerily echoed the Twin Towers' experience inasmuch as numerous rescuers also perished at this Alabama mine. In fact, 12 miners lost their lives in a second explosion while trying to rescue a miner who had been immobilized by an explosion that had happened nearly an hour earlier. Communication problems contributed to the deaths of the 12 rescuers; we believe the rescuers were given insufficient and faulty information about the underground conditions, and attempted the rescue without knowing the hazards they faced.

Quecreek.—In July 2002, 9 miners were trapped by a water inundation in a Pennsylvania mine, after 9 others were able to escape. The trapped miners were rescued 4 days later; again, communication inadequacies frustrated an easier and quicker rescue.

Sago Mine; 12 fatalities.—On January 2, 2006, this tragedy claimed the lives of 12 miners, while the full extent of injuries to the 13th miner, Randal McCloy, Jr., remain uncertain. Hopefully, with time we will learn all of the conditions that caused and contributed to these lost lives.

These dramatic events represent only the headline-grabbing incidents. Thousands of miners are still disabled and dying from black lung disease, while other miners also die in mining accidents each year. Typically they die one or two at a time, from roof falls, equipment failures, and other accidents.

There are also countless near-misses that occur on a regular basis. In fact, just since August 2000, the Mine Safety and Health Administration, known as MSHA, has records of well over 400 mine fires, ignitions, explosions and inundations that far too-easily could have developed into significant disasters and fatalities; many other incidents likely went unreported.

With better regulations, more regular enforcement, and with support from the highest echelons of the Agency, many of these accidents could have been prevented. Senseless deaths and injuries must stop. Mining will probably always be a dangerous job. But we can do a lot more than we are doing today to make it safer. Miners should not have to get sick, or to risk their lives just by going to work.

The most basic point I wish to make today is that as a nation, and as an industry, we already possess the knowledge and the ability to prevent most of the deaths that are still occurring in the coal mining industry. What is needed is a real commitment by our government—in this case, MSHA—to do better.

For example, if MSHA would require additional oxygen units ("self- contained self-rescuers") to be stored on the section where miners work, and throughout the underground mine in addition to the self-rescuers that each miner is required to have, then miners trapped in an emergency would have a better chance of surviving. Self-contained self-rescuers are what miners don to escape, when noxious air fills an underground mine after a fire or explosion. They typically last for about one hour. We understand that miners trapped at Sago wore self-rescuers. However, there is now no MSHA requirement that coal operators store any additional oxygen units in their underground mines. We can only speculate about whether more miners would have survived at Sago if additional self-rescuers had been stored underground. What we do know, is that they would have had a better chance of surviving until help arrived if they had more fresh air to breathe. MSHA should implement a rule requiring additional units to be maintained in strategic locations around the mine in order to provide miners with oxygen from the deepest penetration of the mine out to the surface. No new technology is required to implement this improvement. Self-rescuers can save lives. Are they worth some modest additional costs? We submit the clear answer is "Yes."

There is also technology available today that would enable trapped miners to maintain better communications in an emergency situation. If MSHA would require secondary telephone lines to be placed in a separate entry, that would increase the likelihood that communications could be maintained between miners and those on the surface, even after an explosion or other emergency event. Also, if MSHA would require operators to place walkie-talkies in underground locations, this would facilitate communications during an emergency. Rescue teams rely on walkie-talkies while they travel underground. The equipment is effective for a distance of about 1,000 feet. If trapped miners could communicate with their would-be rescuers, the trapped miners could both provide and receive critical information that could assist in their survival. By not having this ability to communicate, their own rescue is hampered and rescue teams may confront additional hazards that could be avoided if trapped miners would report on what they know. These proposed communications' improvements could be implemented easily; the UMWA urges MSHA to quickly implement these changes. If we can talk to people on the moon, we should be able to talk to those trapped underground.

After the Jim Walters tragedy, and again after the Quecreek near-disaster, the need for better underground communications was crystal clear. Yet, MSHA made no changes to implement any such improvements.

If we could locate trapped underground miners, we could also do a much better job rescuing them. MSHA should draw on all the nation's resources to address this compelling need. We are encouraged by communications we have had with people at the Pentagon about equipment originally developed for aviation security, which might also be applicable to the mining industry. We do know that since the 1970's, there has been electromagnetic technology available that would enable us to locate trapped underground miners. The then-Bureau of Mines reported about this equipment in the 1970's; the equipment was tested and deemed reliable some 30 years ago. Why then aren't we using it today? MSHA should take a look at this electromagnetic technology and all other available technology to see which should be required throughout this industry.

Another way to enhance the chances of survival for trapped miners is to strengthen and expand mine rescue teams. The Mine Act required MSHA to implement regulations ensuring that "mine rescue teams shall be available for rescue and recovery work to each underground coal or other mine in the event of an emergency." 30 U.S.C. §825(s). MSHA's regulation now provides that every operator must establish at least two mine rescue teams, consisting of five members and one alternate, and that these teams be available at all times miners are underground. The regulations also permit an operator "to enter into an arrangement for mine rescue teams" whereby at least two mine rescue teams are available whenever miners are underground. Further, to be deemed "available" the mine rescue teams must be capable of presenting themselves "at the mine site(s) within a reasonable time after notification of an occurrence which might require their services." 30 C.F.R. Part 49. Pursuant to MSHA policy, if a team can arrive in two hours, it is deemed "available." MSHA includes exceptions for small and remote mines, but those exceptions would not have applied to the Sago Mine.

At the Sago Mine on January 2, it was between three and five hours before the first mine rescue teams arrived. Once mine rescue teams arrived on property, including six teams comprised of UMWA members, the sole activity they engaged in revolved around safely extracting the trapped miners. It is extremely important for you to understand that procedures followed during mine rescues are designed to save the lives of trapped miners while also protecting the team members. Rushing into the mine without a complete understanding of its conditions, would be inviting additional problems. You need only review mining history briefly to see examples of rescuers rushing in only to become additional victims. We cannot permit that to occur. Even if underground conditions may not have allowed the rescuers to immediately travel underground at Sago, the five hours' lapse before the first teams arrived constituted valuable time that was simply forfeited. We do not know what arrangements the Sago Mine had to address its mine rescue team obligations, but we do know that a five hour lag before mine rescue teams arrive is unacceptable.

The UMWA submits that every underground coal mine should have mine rescue capabilities on site. These team members should be employees at the facility who would be acutely familiar with the mine. These individuals would not only be best able to carry out many of the duties required in these situations, but would also be uniquely qualified to brief additional offsite teams that may be necessary to complete the rescue. For even small and remote mines, MSHA should require mine rescue teams to be ready when disasters strike. No trapped miners should ever again have to wait three to five hours for rescue efforts to begin.

While we all share the hope that all mining accidents will cease, all of our hopes probably won't prevent more accidents from happening. What we can do, and what we should do, is to give miners their best chance at surviving a mine emergency. I have just described a number of improvements that would be easy to implement; they would also go a long way to furthering miners' ability to survive disasters.

However, unless MSHA requires that these improvements be implemented throughout the industry, miners will continue to face emergencies from a position of compromise.

If the Sago miners had been able to communicate with persons on the surface and the mine rescue teams underground, those miners might have been successfully lead to fresh air. If the location of those trapped miners could have been more-quickly determined, we could have enhanced the possibility of their rescue. These "only-ifs" are too late to change what happened at Sago, but it would be inexcusable if our inaction now would contribute to any more deaths.

There was a national outcry when 78 miners died in West Virginia in 1968. Unfortunately, after the Jim Walters' disaster and the recommendations made from the lessons learned there, no lasting improvements to miners' health and safety came after those 13 miners perished. We cannot allow that to happen again. We need to make real changes so that another tragedy like what happened at Jim Walters and what happened at Sago will not happen again.

MSHA knows how to do better. The Agency itself has performed countless internal reviews and self-analyses; the Federal Government's watchdog agency, the GAO, has given it direction, and the UMWA has communicated both formally and informally about how MSHA can and must do better.

GAO recently focused on shortcomings in MSHA's performance with regard to the underground coal industry. The underground coal industry is the same part of MSHA's jurisdiction that was at issue in all three incidents I highlight in this testimony: Jim Walters, Quecreek, and Sago.

GAO issued its report in September 2003, two years after the Jim Walters tragedy. In its report, which I attach to my testimony, GAO noted that MSHA headquarters was not performing adequately in several key areas. Specifically, the GAO found MSHA failed to ensure violations cited to mine operators were corrected in a timely fashion. In fact, GAO found that of all the citations issued by the Agency, including those written as "significant and substantial," despite inspector-imposed deadlines by which problems were to be abated, 48 percent of the time the Agency failed to follow-up in a timely fashion to see if the operator fixed the hazards.

GAO also found that MSHA collected information about accidents and investigations, but then failed to use the information effectively to prevent future accidents. GAO noted that MSHA does not even collect information about how many hours contractors work at mine sites, making it impossible to compute the fatality and injury rates for particular mines. It further found that MSHA failed to ensure that the ventilation and roof control plans are reviewed every six months, even though the Mine Act and applicable regulations, as well as MSHA's long- standing policies, require that these reviews occur on a semi-annual basis.

After MSHA completed its investigation into the Jim Walters disaster, the Agency also performed an Internal Review of MSHA's actions before the explosions to "improve our inspection process to better protect our nation's miners." The review compared what MSHA actually did with what the Mine Act requires it to do. A number of problems were identified as deficiencies "at both the district and headquarters level", deficiencies "relevant to inspection procedures, level of enforcement, plan reviews, the [Alternative Case Resolution Initiative] and accountability programs, supervision and management, and headquarters oversight."

Unfortunately the Agency's top managers have done little to move any of the necessary improvements from recommendation to reality. We hope that by having Congress add its voice now, along with the public's demand for its better performance on the heels of Sago, MSHA will finally re-focus its attention.

MSHA should promulgate regulations that meaningfully improve miners' health and safety, and MSHA must consistently enforce the regulations it has. The circumstances that played out on January 2, 2006 have become all too familiar. The events that unfold at these mine disasters change very little, they focus attention for a brief moment but, as the spotlight fades, MSHA is content to ignore demands for change. We cannot and will not permit that to happen here.

The Mine Act is an important piece of legislation. I am not here to advocate an opening of any part of that historic and effective law. The UMWA is here to simply demand that it be enforced by MSHA as Congress intended. The real problems that exists with the current application of the Act lie in an administrative malaise. Instead of enforcing it aggressively, the Department of Labor has been whittling away at the Mine Act. Too often MSHA relies on "policies," which are developed internally and without public comment, to circumvent the Mine Act. This reduction in MSHA's effectiveness didn't happen overnight; it has been a problem for much too long. We have been critical of MSHA under both Democratic and Republican administrations. But now we call upon Congress to put a stop to MSHA's inadequacies, to turn the

Agency around so it can enforce the Act the way Congress intended in 1969, and again in 1977.

To make the needed changes will require MSHA to take a new direction, beginning at the top. MSHA needs to have a larger budget for coal enforcement. The Agency spends too much effort at "compliance assistance," and too little on enforcement. It needs to bolster its expertise. MSHA has some excellent mine health and safety experts working in field offices throughout the Country. Yet, they have not been receiving support from those above them. Too often, an inspector will write citations and orders upon finding violations of regulations, only to have them compromised away through conferences and settlements. While MSHA employs many experienced and dedicated inspectors, there is a failure on the part of individuals sitting at the higher levels of power to support the efforts of the MSHA field staff to enforce the Mine Act.

MSHA's failure to aggressively enforce the Mine Act is the result of many factors. MSHA is full of former mine management executives who spend too much time trying to appease their friends, and too little time looking out for miners' interests. How can we expect a regulatory Agency to effectively and fairly carry out its duties and responsibilities when it is run by some of the very coal operators it was designed to regulate? Many of MSHA's top-level administrators spent years opposing any regulatory efforts attempted by the Agency. They continue to be influenced by other coal operators, effectively muting the voices of miners who need a healthy and vibrant enforcement Agency.

For years, the fox has been inside the henhouse at MSHA. During his tenure as Assistant Secretary for MSHA, David Lauriski, scrapped 18 proposed rules on topics MSHA had already identified as needing attention. I attach to this testimony a list of those proposed rules that former-Administrator Lauriski withdrew, and will discuss the importance of a few of them. However, withdrawing proposed rules was not the extent of his actions at the Agency. He also pushed to completion several regulations that weakened health and safety, because they benefitted mine operators. For instance, under his watch MSHA implemented a regulation that allows mine operators to ventilate areas of the mine where miners work with air that has already passed through the conveyor belt entry, a practice Congress specifically prohibited in the Mine Act (30 U.S.C. §863 (y)); and under his leadership MSHA implemented a regulation that permits the use of diesel-powered generators in confined areas of underground mines. While these practices may increase production for mine operators, they also pose new and significant risks to miners.

Unless something changes soon, history will repeat itself at the Agency. Later this month, on January 31, 2006, a hearing is scheduled to consider the Administration's nominee for Assistant Secretary of Labor for MSHA. This second nominee by this Administration has been a coal mine supervisor for most of his life. His lone excursion into health and safety was marked by repeated attempts to limit regulations and reduce the health and safety protections of miners in the Commonwealth of Pennsylvania. MSHA—and our nations' miners—cannot afford having another coal operator sitting as the Assistant Secretary. Miners deserve an advocate and an ally, not another coal boss. There should be no doubt that when top-level appointees are too cozy with the industry, miners ultimately pay the price.

Among the regulations MSHA stopped developing, and withdrew from its agenda were those addressing coal dust exposure; self-rescuers; mine rescue teams; accident investigations; and training/re-training.

The proposed rule on *Occupational Exposure to Coal Mine Dust* was withdrawn on September 4, 2002. That proposed rule was drafted to comply with the Secretary of Labor's Advisory Committee on the Elimination of Pneumoconiosis Among Coal Mine Workers, and was intended to decrease the level of respirable coal dust miners could be exposed to during a working shift. By cutting the permissible exposure level in half, miners would be less likely to contract debilitating black lung disease. Application of such a standard would also have significantly reduced the amounts of highly-explosive float coal dust released into the mine atmosphere. Such a regulation would have significant health and safety benefits for underground miners. Unfortunately, the only efforts regarding coal dust that MSHA made under former Assistant-Secretary Lauriski was a proposal that would have allowed respirable dust levels to increase by four fold. To put it mildly, his proposal was not well-received and he ultimately withdrew it.

MSHA's proposed rule on *Underground Coal Mining; Self-Contained Self-Rescuer* was withdrawn on September 24, 2001. Self-contained self-rescuers have not been updated to keep up with technology. They were first required in 1981 and little has changed since then. Some of these devices were found to be inoperable for a variety of reasons including deteriorating hoses, contaminated chemical beds, and unrealistically long shelf lives being approved by MSHA. The industry has also been plagued with the fact that miners sometimes cannot properly don the units in emergency situations. Moreover, with MSHA's continued acceptance of the status quo, technological advances of these breathing devices is being stymied. In the legislative history of the Mine Act, Congress indicated that mining regulations should be technology-driving, to maximize miners' protections. We had hoped that with the promulgation of a new rule addressing self-rescuers, the existing problems would be addressed, and technological advances encouraged. The UMWA is convinced that such a rule would have been the catalyst for a new generation of self-rescuer devices. While operators are willing to invest in new technology when it increases production, it appears that they are not so willing to invest when in miners' health and safety.

The proposed rule on *Mine Rescue Teams* was withdrawn on September 4, 2002. The basis for moving this rule forward is quite simple: the UMWA and many industry officials recognize that, with mining operations contracting in the late 1980's through the 1990's, the number of mine rescue teams was disproportionally reduced. This left large coverage gaps. The industry is also facing an overall aging of the workforce, and this also adversely impacts participation in those rescue teams that remain active. In May of 2000, when it published this pre-rule, MSHA stated: "We are assessing the current regulations to identify problem areas where we might increase flexibility and increase safety for miners." However, instead of promulgating a rule that would improve rescue teams' availability and capabilities, MSHA eliminated further work on rescue teams regulations. Meanwhile, it permits operators to expand on the ill-advised practice of contracting out such work. Withdrawing the proposed rule effectively eliminated any meaningful improvement in comprehensive mine rescue activity, but it also afforded some mine operators the opportunity to disband teams so they could increase their profits.

On August 16, 2001, the Agency withdrew its proposed rule on *Accident Investigation Hearing Procedures*. MSHA has no formal rules for conducting investigations. While it has established policies, the investigation process is subject to change on the whim of the individuals running the Agency. This is exactly what developed in the early stages of the investigation of the Sago Mine disaster. Not only do questions arise about who should participate in various parts of the investigation, but for the Jim Walters' investigation, for example, MSHA did not conduct a single public hearing to ensure that all relevant information was presented.

On September 24, 2001, MSHA withdrew its proposed rule on *Training and Retaining of Miners*. This critical proposal would have increased the number of hours operators are required to set aside annually for health and safety training of miners. This training includes first aid, donning and using self-contained self-rescue devices, fire drills, and exiting the mine in the event of an emergency. Miners' lives may turn on the quality of their training. The training must be updated and improved. There should be no doubt that the better trained miners are, the better equipped they will be to escape a mine emergency.

These are only five examples of the18 regulations the Agency, under this Administration, determined to be insignificant, too burdensome, or too costly to promulgate. Several of them may have had significance in what developed at the Sago Mine. We may never learn that for a certainty, but now is the time to require better rules, offering better protections so that miners have a better chance of surviving mine emergencies.

Congress can, and should demand that MSHA do in 2006 all that Congress demanded in 1969 and again in 1977. Regulations that were in the pipeline in 2001 and 2002 should be reactivated and finalized in a timely fashion; regulations that are already in place must be enforced regularly and aggressively. Now that the spotlight is on the issue of miners' health and safety, we have a unique opportunity to make improvements.

The status quo is inadequate. The government failed the Sago miners, and when it failed them it failed all miners. In enacting the Mine Act, Congress plainly stated: "Congress declares that (a) the first priority of all in the coal or other mining industry must be the health and safety of its most precious resource the miner." (30 U.S.C. §801.) We take that admonition seriously; everyone else associated with the mining industry must re-establish miners' health and safety as their top priority, too.

I thank you for your interest in miners' safety and would be happy to answer your questions.

Senator SPECTER. Thank you, very much Mr. Roberts. I might say at this point, if there are other key factors you want to call to the attention of the subcommittee which will not be covered in the

questions session, we would welcome your supplemental responses
if they're not covered in any written testimony which has been submitted. Attachments to your testimony will be in the committee
files.

We now turn to Mr. Ben Hatfield, President and CEO of International Coal Group. Prior to joining the company, Mr. Hatfield
held leadership positions with the Eastern Operations of Arch Coal,
El Paso's Energies Coastal Coal Company, and Massey Energy
Company, Bachelor of Science, and Mining Engineering from Virginia Tech. Thank you for coming in today Mr. Hatfield, and the
floor is yours.

STATEMENT OF BENNETT K. HATFIELD, PRESIDENT AND CEO, INTERNATIONAL COAL GROUP, ASHLAND, KENTUCKY

Mr. HATFIELD. Thank you Mr. Chairman and members of the
subcommittee. I appreciate the opportunity, and invitation to testify here today. First, I want to extend my deepest sympathies to
the families whose loved ones were lost at the Sago Mine. Our communities continue to mourn the deaths of our friends and coworkers. We also pray for the full and speedy recovery of Randy McCloy,
our lone survivor.

Second, I commend the heroic efforts made by many mine rescue
teams from other companies, and volunteers who came forward
during our community's time of need. The outpouring of support
from the Buckhannon area communities has overwhelmingly demonstrated how West Virginians come together in times of crisis. We
also appreciate the support provided by the Federal Mine Safety
and Health Administration, and the State of West Virginia.

Our company is working closely with MSHA and the State in the
formal investigation to determine the cause of this tragic accident.
I personally promise to use all available resources to get the answers that the friends, families, and coworkers need and deserve.

My written statement details the events of January 2 through 4,
beginning with an explosion at about 6:30 a.m. in the midst of a
violent lightening storm. This explosion in an instant separated
brother from brother, uncle from nephew, and friends from friends.
The supervisors onsite immediately reacted with a rescue effort in
an attempt to reach their fellow miners. Sadly, their efforts were
to no avail. It would be many more hours before conditions improved to allow additional rescue teams to enter the mine, and
throughout we waited, waited, and then waited more.

During those dark hours we prayed for 13 miracles, and in the
end, were blessed to receive one. We deeply regret the pain caused
by that early and erroneous communication to the families. There
was never any intent to misinform, mislead, or raise false hopes.
We, like you, and like the families of the miners, and like Americans throughout our country road that emotional roller coaster that
has sadly lead us to where we are today.

Within the next few days, investigators should be able to safely
enter the mine to begin determining the cause of the accident.
After a devastating accident of this nature, it is understandable
that questions are asked about the Sago Mine's safety record. My
written statement details some of the safety improvements made
by our company beginning on June 1, 2005, when we assumed

management oversight. What is important to emphasize is that since June 1, our company has worked closely with Federal and State Regulators to make this mine as safe as possible. For example, we were the first coal company to voluntarily invite MSHA's technical support group on incident reduction, to help implement a new program to continually improve mine safety. Most importantly, that effort helped us to dramatically reduce the injury rate at the Sago Mine by nearly 60 percent from the first half of 2005 to the second.

PREPARED STATEMENT

I believe that if there is any good that can come of this horrible event, it will be in inspiring greater innovations in mine safety. Although it's too early to determine the cause of the tragedy, we intend to be a leader in efforts to prevent future tragedies. As an example, our company is already moving forward with evaluation of improved wireless communication technology, and supplemental oxygen supplies on demand. We will work on our own, and with others in the mining community to improve technology, and we will continue to base our business decisions on worker safety as the first, and most critical consideration.

We must learn lessons from this explosion that will better protect coal miners. That is our company's commitment to the families of the 12 miners who perished. Thank you.

[The statement follows:]

PREPARED STATEMENT OF BENNETT K. HATFIELD

Mr. Chairman and members of the Subcommittee, I am Ben Hatfield, President and Chief Executive Officer of International Coal Group, Inc. (ICG). By way of brief background, International Coal Group, Inc. is a leading producer of coal with operations in West Virginia, Kentucky, Maryland, and Illinois. We employ approximately 2,100 people throughout our operating area. During 2005, we sold about 19 million tons of coal to utility, industrial, and metallurgical customers located throughout the Eastern U.S. I appreciate the invitation to testify today.

First, I want to extend my deepest sympathies to the families whose loved ones were lost at the Sago Mine. Our community will continue to mourn the deaths of our friends and coworkers. We also pray for the full and speedy recovery of Randal McCloy, Jr.

Second, I commend the heroic efforts made by many mine rescue teams from other companies and volunteers who came forward during our communities' time of need. The outpouring of support from the Buckhannon area communities, churches, local businesses, civic organizations, and emergency personnel has overwhelmingly demonstrated how West Virginians come together in times of crisis. We also appreciate the support provided by the Federal Mine Safety and Health Administration (MSHA) and the State of West Virginia.

Our Company is working closely with MSHA and the State of West Virginia in the formal investigation to determine the cause of this tragic accident. We vow to use all available resources to get the answers that the families, friends, and coworkers need and deserve.

BRIEF SUMMARY OF EVENTS

I would like to briefly summarize what we know about the events of January 2, 3 and 4, 2006. At approximately 6:00 A.M. on Monday morning, January 2, after their travel route and worksites had been reported safe by certified safety examiners, two production crews and mine support staff totaling 27 miners entered the Mine. Each production crew needed to travel about two miles to reach the two working sections (First Left and Second Left) of the Mine where they were scheduled to work. The Second Left crew entered the Mine on a rail manbus that departed roughly 10 minutes ahead of a similar manbus carrying the First Left crew. One of the certified safety examiners had remained underground and traveled to his normal workstation. Therefore, a total of 28 miners were underground.

At 6:31 A.M., Sago mine management heard the audible alarm of the mine monitoring system indicating the presence of carbon monoxide underground. At about the same time, the electrical power supply to the Sago Mine was disrupted. All of this occurred in the midst of a violent storm with unusually strong lightning strikes. Shortly thereafter, the supervisor of the First Left production crew telephoned the dispatcher on the surface to report that his crew had just experienced a very strong rush of air with substantial smoke and dust emanating from deeper in the Mine. Mine management directed the First Left supervisor to bring his crew out of the Mine through one of the two primary escapeways. No communication was received from the Second Left crew of miners. Repeated efforts by mine management and the First Left crew to contact the Second Left crew, via mine phone and underground walkie-talkie, were unsuccessful. So, mine management immediately became concerned that they were in danger.

At 6:41 A.M., Mine Superintendent Jeff Toler, whose uncle was one of the missing miners, and three other mine supervisors headed underground to investigate. After traveling about one and a half miles by rail manbus, they encountered the First Left crew coming out of the Mine on foot. The rail manbus was given to the First Left crew to expedite their safe exit from the Mine. The supervisor for the First Left crew, whose brother was one of the missing miners, joined the mine management team in an attempt to reach the Second Left crew located roughly 2,000 feet deeper in the Mine. They quickly gathered tools and ventilating materials and then proceeded toward the Second Left section. This initial rescue effort by the five-man management team continued for over two hours as the group encountered thick, black smoke and attempted to redirect ventilating air to open a route of access to the missing crew. Repeated calls to the Second Left crew via mine phone received no response. The rescuers became increasingly concerned that a possible explosion could be ignited as they directed fresh air toward the Second Left section. Consequently, the mine management group exited the Mine at 9:45 A.M.

Meanwhile on the mine surface, at about 7:00 A.M., company safety managers not already on site were called and briefed on events at the mine. Following various communications between those safety managers and mine management on site regarding immediate emergency procedures required, we began calling MSHA and State safety officials to report the accident at about 7:40 A.M. Both MSHA and State safety officials were reached between 7:56 and 8:28 A.M., and began arriving on site soon thereafter. At 8:32 A.M., MSHA inspector Jim Satterfield orally implemented an emergency mine closure order (a "103k order") prohibiting further entry to the Mine. At about the same time, State mine inspectors began monitoring the air quality at the Mine portal. High concentrations of carbon monoxide were found, indicating significant risk of an active underground mine fire that could ignite an explosion, so state and Federal mine regulators on site determined it was not safe for mine rescue teams to enter. This agonizing process of monitoring the carbon monoxide and methane levels in the Mine air and having to wait and wait for confirmation that it was safe to enter would continue throughout the day.

In anticipation that mine rescue teams were going to be allowed to enter the Mine soon, the first call to the Barbour County Mine Rescue Team was made at 8:04 A.M. That Team arrived on site at approximately 10:40 A.M., and waited for State and Federal authorities to approve their entry into the Mine. Other mine rescue teams were also contacted, and continued to arrive at the Mine through the course of the day, with eventual deployment of 13 to 15 mine rescue teams by late afternoon on January 2.

After experts from MSHA, the State of West Virginia, and the Company agreed that an underground mine fire was no longer likely based on the air monitoring results, the first mine rescue team entered the Mine (carrying special breathing apparatus) at 5:51 P.M. The search and rescue efforts continued throughout January 2 and January 3. Progress had to be careful and deliberate to protect the safety of rescuers, given that many rescuers had fallen victim to secondary explosions in coal mine disasters of years past. During the evening of January 3, a rescue team found one miner's body. Just before midnight, our remaining 12 missing miners were found. As a result of the extreme difficulties in communication hundreds of feet below the surface while wearing special breathing apparatus, the now well-known miscommunication about the number of survivors occurred. We, too, rode that same emotional rollercoaster and suffered the inevitable pain when the truth was learned.

We expect that investigators will be able to safely get back into the Mine soon to determine the cause of the accident.

SAFETY AT THE SAGO MINE

Even before we completed the acquisition of the Sago Mine, on November 18, 2005, we assumed management oversight through a consulting agreement that allowed us to begin making safety improvements as of June 1, 2005. Since that time, our Company has worked closely with Federal and State regulators in an effort to make this Mine as safe as possible. Specifically ICG has voluntarily:

—Rehabilitated two miles of primary intake escapeway and more than doubled the amount of fresh air reaching the working sections. This is the escapeway used by the surviving crew.

—Upgraded the rail system used to move miners and supplies into and out of the Mine.

—Invited MSHA's Technical Support Group on Incident Reduction to help implement a new program to continually improve mine safety. We were told that ICG was the first coal company to voluntarily work with MSHA under the Agency's Incident Reduction Program.

—Required Sago Mine hourly employees to receive eight hours of supplemental safety training during September 2005, in addition to the extensive training already required under the Mine Act. Then, during October-December, we required our Northern West Virginia Region supervisors to receive two days of supplemental training at the MSHA Training Academy at Beckley, West Virginia.

—Established a Performance Group Initiative that gives every employee a forum for addressing any safety or production concerns or suggestions, anonymously, if they so chose, and addressed any points raised in monthly meetings.

These voluntary initiatives helped us to dramatically reduce the lost time injury rate at the Sago Mine by nearly 60 percent from the first half of 2005 to the second.

Sago's employees are well-trained, skilled coal miners who understand safety. Each employee is aware that if an unsafe condition is identified, they are authorized to withdraw immediately from the hazardous area and notify their supervisor of the danger.

MSHA data shows that:

—Mining operations at the Sago Mine more than doubled between 2004 and 2005, prompting MSHA to dramatically increase—by 84 percent—its on-site inspection and enforcement presence.

—Of the 208 citations, orders, and safeguards issued in 2005, none involved an immediate risk of injury and all but three had been fully corrected by January 2. The three remaining issues, which relate to roof control, are being addressed by Sago in compliance with the Mine Act.

—Only when MSHA completes its investigation will we know the cause of the accident, but we do know that none of the health and safety violations cited by MSHA at Sago Mine last year involved immediate risk of injury and that the Mine has worked to correct all health and safety problems in accordance with the requirements of the Mine Act.

The Mine Act also authorized MSHA to shut down an operation that is unsafe, and MSHA's trained mine safety professionals, who were at the Mine nearly every working day in the several months before the accident, would certainly not have allowed the continued operation of the Sago Mine if they believed it to be unsafe. In addition, as required by law, our certified mine examiners inspect the Mine before and during every shift. They, too, are fully authorized to shut down any part of the Mine they consider unsafe. While the tragic events of January 2 confirm that we must be ever vigilant on mine safety, the safety record at the Sago Mine demonstrates that our management team aggressively focused on mine safety and protecting our people.

THE FUTURE OF MINE SAFETY

Although it's far too early to determine the cause of the tragedy and the extent to which it may have been preventable, we intend to be a leader in the effort to identify and develop safety technologies that will help to prevent future tragedies. We will work on our own, and with others in the mining community, to improve technology, and we will continue to base our business decisions on worker safety as the first and most crucial consideration.

We expect that this terrible series of events will further motivate the entire mining community to identify and implement significant improvements in mine safety through cooperation, information sharing, and improved technologies. For example, working with MSHA, we should vigorously seek to advance the development of permissible wireless communications and breathing apparatus technologies that could further improve the coal industry's underground mine rescue capabilities. Also, this experience highlighted critical weaknesses in the design of MSHA's V–2 robot that could likely be remedied with the technology now used by NASA in space exploration. Once the actual cause of the Sago explosion is known, there may be more specific measures that could help prevent a recurrence.

We must learn lessons from this explosion that will better protect coal miners. That is our Company's commitment to the families of the 12 miners who perished.

Senator SPECTER. Thank you very much Mr. Hatfield. Our next witness is Mr. Chris Hamilton, Senior Vice President to the West Virginia Coal Association, an organization which he has been affiliated with for more than 20 years, has a Mine Foreman Certifications from West Virginia to Ohio, Undergraduate and Graduate Degrees from West Virginia University. I thank you for joining us Mr. Hamilton, and we look forward to your testimony.

STATEMENT OF CHRIS R. HAMILTON, SENIOR VICE PRESIDENT, WEST VIRGINIA COAL ASSOCIATION, CHARLESTON, WEST VIRGINIA

Mr. HAMILTON. Mr. Chairman, Senator Byrd, other distinguished members of the committee, thank you for the invitation to address this committee, and for placing this important topic on your agenda for review and discussion. In West Virginia, we have approximately 40,000 individuals, men and women who work directly in, or around a coal mining facility, and without exception all miners, managers, engineers, support staff, along with our entire State, have been deeply saddened by the Sago and Alma tragedies, and the mourning will continue for years to follow.

Our hearts and prayers are with the families and loved ones of the miners who perished at Sago and Alma, and we continue to pray for Randal McCloy's full and speedy recovery. We now extend those prayers to the families of Don Bragg and Ellery Hatfield of the Alma Mine tragedy as well.

West Virginians share a special bond with their families, church, and communities. They have an unparalleled inner strength, inner faith, and nowhere is that bond more prominent than in West Virginia's coal industry. For the record, the West Virginia Coal Association wholeheartedly embraces Governor Manchin in his plain spoken sentiments, that no miner should ever be fatally injured in a West Virginia coal mine. We also fully support the Governor's commitment to operate the safest mines in the world. We will commit the necessary resources over the months to come, and we'll do everything humanly possible to achieve that shared goal.

First and foremost, that is our commitment which we believe is realistic and achievable. We also maintain that the primary responsibility for achieving that goal rests firmly with those who own, operate, and manage coal mining operations. A responsibility we not only acknowledge, but aim to fulfill. These tragic events have caught the eye of practically all of America in the past three weeks, and the media has done a good job, an accurate portrayal of the courage and overall character of the men and women who have selected mining as a profession. They have a passion for their work, and they do it with great pride, in an exceptional level of professionalism.

Unfortunately, the events of January 2 and those of last week, have not accurately portrayed how technologically advanced mining has become, and all the progress, and safety achievement that's been made over the past several decades. But, one mining death is too many, and despite all this progress recorded in recent years, we now realize that much work remains. Particular focus is required in the post accident phase, so that the effect of accidents can be truly minimized or mitigated.

By its nature, mining is unique unlike any other business or industry, and that it is dependent on natural conditions, and natural geology. Through their skills, training, dedication, and hard work, miners attempt to control and manage the challenges of their environment, and they're good at it. It requires a supreme vigilance every minute of every shift. New mining technologies such as long wall mining, remote controlled equipment design, and mine wide atmospheric monitoring systems, combined with the extraordinary skill and experience level of today's workforce has truly lead to safer conditions, and fewer accidents.

As a relevant part of my testimony on record today, I incorporate a copy of the most recent directory of mines, which is published annually by the West Virginia Office of Miners Health Safety and Training. It contains useful statistical information, and also charged the Mine Safety Performance of the industry over the years. This directory reflects a dramatic reduction in mine related deaths since the passage of the 1969 Act, from 162 fatal accidents reported, to just three for all of 2005. Again despite all this progress, we realized that much work remains, and that vigilance will be continued day in, and day out.

This report also depicts a significant reduction of mine accidents, and lost time injuries over the same period. The State's annual report also reveals that the State of West Virginia has one of the more comprehensive mine safety programs found anywhere in the country, with a full compliment of mine safety inspectors, safety officials, and extremely aggressive legislative and regulatory program. I'm pleased to report that the State of West Virginia is taking a lead in dealing with this terrible tragedy that we've all experienced. I believe we will learn later this afternoon that three important pieces of legislation will be introduced into the West Virginia legislature for immediate action today. I believe that legislation will address the points raised by Senator Byrd in his opening remarks. I think we'll see legislation calling for a central rapid response program, and modeled after our State Office of Emergency Services. I think we'll see requirements calling for additional self contained breathing apparatuses stored at strategic locations underground, and I think we'll see legislation that preempts, and that drives technological advances to have improved communication systems within the mine itself, and from above the surface to the underground mine, and connected with all miners. I'm convinced we'll see that. I'm confident we'll see that legislation introduced today, and it's my understanding that by mid afternoon, that legislative action ought to be well on its way.

We have joined with the Governor, we want to be part of the solution, we recognize despite all of the improvement, all the technological advancements, that much work remains. We are truly here, part of this proceeding to offer our help and our assistance to my colleagues sitting at this table, to the State of West Virginia, and to this particular Subcommittee. We have a wealth of technical capability.

Senator SPECTER. Mr. Hamilton, you're substantially over. Could you summarize at this point?

PREPARED STATEMENT

Mr. HAMILTON. Yes. Again, I can't express enough the sorrow that is shared by all West Virginians. We compliment you Mr. Chairman, other senators for placing this on your agenda, and we look forward to working with you in the months to come, to address these concerns. Thank you.

[The statement follows:]

PREPARED STATEMENT OF CHRIS R. HAMILTON

Mr. Chairman, Members of the Committee: Thank you for the invitation to address this Committee and for placing this important topic "Coal Mine Health & Safety" on your agenda for review and discussion.

INTRODUCTION

My role and contribution to today's hearing will be defined by the following four (4) key points: First, to express our heartfelt prayers for the families who suffered great personal loss at the Sago Mine. Our prayers continue for Randal McCloy's full recovery and for his wife and family. We now extend those prayers and our state's unique circle of support to the families of Don Bragg and Ellery Hatfield of the Aracoma Mine tragedy. The deceased miners will forever be with us as we implement the necessary steps to improve coal mine safety and prevent recurrences. We also thank the mine rescue team members, the State, Federal, and company officials who directed and guided their heroic and brave efforts at Sago, and whose performance in those dark and anxious hours will be analyzed for years to come. It is our hope that their performance will be constructively reviewed with an eye towards improving future rescue efforts; Thirdly, we are here as one of the nation's largest trade associations to offer our pledge to work with you in whatever capacity you deem appropriate in the discharge of your important work and to direct our Association's collective attention towards the identification and implementation of appropriate remedial measures; and Fourth, and subordinate to the preceding points, is the perceived need to preserve the integrity and future of the coal industry—to implement the necessary changes from the lessons learned from the horrific accident that brings us here today and to elevate the understanding and appreciation of our industry which means so much to West Virginia and to our nation!

My personal background: I have nearly thirty-five years of experience in the coal mining industry beginning in 1971 during the immediate implementation of the 1969 Federal Mine Health and Safety Act and over thirty years of experience in mine health and safety.

I worked as an underground miner and for underground and surface mining companies. I have also worked for the Federal and State mine safety agencies as a mine safety professional and safety instructor—certified to train and certify miners in all aspects of mining and mine safety including mine emergency preparedness and mine rescue operations.

As Training Director for the West Virginia Department of Mines (for then Governor Jay Rockefeller), I was responsible for approving mine training facilities, mine training plans and individual mine training instructors.

I possess underground Mine Foreman—Fire Boss certifications from WV and the state of Ohio where I worked for several years in the industry. I received my undergraduate and graduate degree from West Virginia University and have also completed many college level courses in mine safety, mining technology and mine industrial engineering.

I presently serve under gubernatorial appointment on the West Virginia Coal Mine Health and Safety Board; the West Virginia Mine Safety and Technical Review Committee; The West Virginia Board of Miner Training Education & Certification; and, the West Virginia Diesel Equipment Commission.

During my tenure on as a mine safety official, I have been involved in the review/investigation of serious mining accidents and practically every single mining death in West Virginia for the past twenty-five years.

As a member of the West Virginia Board of Coal Mine Health & Safety (the only independent entity in West Virginia with a statutory charge to investigate and respond to mine accidents), I will be part of the state's investigation and regulatory response to the Sago and Aracoma accidents!

West Virginia's coal industry is comprised of approximately 40,000 individuals who work directly in, or around a coal mining facility and without exception, miners, managers, engineers and support staff along with our entire state have been deeply saddened by the "Sago and Aracoma tragedies" and will continue to mourn for years to follow.

Our hearts and prayers are with the families and loved ones of the miners who perished in the Sago incident and we continue to pray for Randal McCloy's full recovery. We now extend those prayers and our state's unique circle of support to the families of Don Bragg and Ellery Hatfield of the Aracoma mine tragedy. I would observe that next to the immediate families of the deceased miners, nobody is saddened more than mine management officials over this tremendous loss. West Virginians share a special bond with their families, church and communities.

They have an unparalleled inner strength and inner faith and no where is that bond more prominent than in the coal industry.

For the record, The West Virginia Coal Association wholeheartedly embraces Governor Manchin's sentiments "that no miner should ever be fatally injured in a West Virginia coal mine". We also fully support the Governor's commitment to operate the safest mines in the world! We will commit the necessary resources over the months to come and will do everything humanly possible to achieve that shared goal!

First and foremost, that is our commitment which we believe is realistic and achievable!

We also maintain that the primary responsibility for achieving that goal rests firmly with those who own, operate and manage coal mining operations. A responsibility we not only acknowledge but aim to fulfill!

These tragic events have caught the eye of practically all of America in the past three weeks and the media has presented an accurate portrayal of the courage and overall character of the men and woman who have selected mining as a profession. They have a passion for their work and they do it with great pride and an exceptional level of professionalism!

Unfortunately, the events of January 2nd and those of last week have not accurately portrayed how technologically advanced mining has become and all of the progress and safety achievement that's been made over the past several decades. But one mining death is one too many and despite all the progress recorded in recent years, we now realize that much work remains! Particular focus is required in the post accident phase so that the effect of an accident can be minimized or mitigated!

By its very nature, mining is unique (unlike any other business or industry) in that it is dependent on natural conditions and geology. Through their skills, training and hard work, miners attempt to control and manage the challenges of their environment—and they are good at it! It requires a supreme vigilance every minute of every shift.

Undoubtedly coal mining is a dangerous occupation with unique hazards inherent to the workplace but I would maintain that mining is much safer today than what was realistically believed possible a few short years ago.

New mining technologies such as longwall mining systems, remote-controlled equipment design and mine wide atmospheric monitoring systems combined with the extraordinary skill & experience level of today's workforce has led to safer conditions and fewer accidents.

As a relevant part of my testimony and record today, I incorporate a copy of the most recent "Directory of Mines" which is published annually by the West Virginia Office of Miners' Health, Safety & Training. It contains useful statistical information and charts the mine safety performance of the industry over the years.

The "Directory" reflects a dramatic reduction in mining related deaths since passage of the 1969 Mine Safety Act when 162 fatal accidents were recorded to 3 for all of 2005. It also depicts a significant reduction in mine accidents and lost time injuries over this same period.

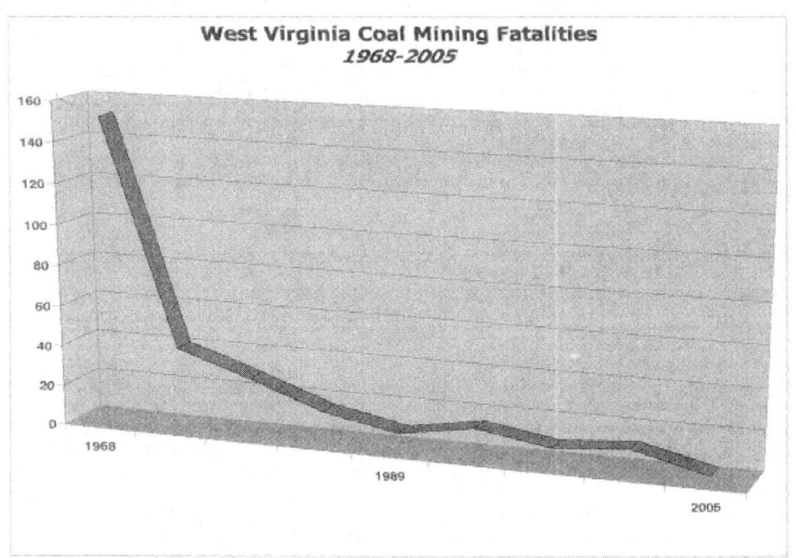

The state's annual report also reveals that the state of West Virginia has one of the more comprehensive mine safety programs found anywhere in the country with a full complement of mine safety inspectors, safety officials and an extremely aggressive legislative and regulatory program. It is also noteworthy to point out that no provision exists under Federal law for States to acquire "primacy" over the administration of mine safety laws.

Consequently, all West Virginia mines are examined by State and Federal inspectors throughout each and every quarter.

The significance of the industry and the important role coal plays in our everyday lives, which ranges from our basic quality of life to national defense and national security and should also serve as a tribute to the men and families of Sago!

Over the past several weeks we have heard local, regional, national and international media sources all ask a similar question: Why do we continue to mine coal?

Coal, and in particular, West Virginia coal, is crucial to our advanced society and extraordinarily quality of life. Coal continues to account for over fifty percent of the America's electricity. In West Virginia that figure is closer to 99 percent.

Over the past several decades our state's coal industry has a remarkably record of safety achievement, reclamation accomplishments and environmental stewardship. We are coordinating proposed mine sites with local and state planning agencies to ensure meaningful and more productive development occurs.

West Virginia is a shining example of where you can have a robust coal industry along with a thriving tourism industry—you can truly have both and I submit to you that nobody is doing it better!

Today's industry represents a technologically advanced enterprise with a highly skilled and efficient workforce and has established a healthy presence in an international marketplace.

West Virginia produces approximately 160 millions tons of coal annually. Of that total, over 105 million tons or 65 percent percent comes from underground mines and approximately 55 million tons of coal is produced from surface mines.

West Virginia continues to lead the nation in underground coal production and is second only to the state of Wyoming in overall coal production. West Virginia is the world's leader in Longwall mining and is the leading coal export state.

All told, West Virginia coal is shipped to 23 foreign countries and accounts for approximately one half of the United States total export product leaving domestic boundaries contributing immensely to the United States balance of trade.

We also have more processing plants than any other state, more transportation outlets and one of the more elaborate transportation systems and infrastructures you find anywhere in the world. It is comprised of rails, trucks and barges and we have the best quality and variety of coals found anywhere in the world.

Due to its clean and high quality, West Virginia coal is shipped throughout the eastern half of the United States to thirty-three states to generate electricity for industrial and household energy and for coking and steel production. West Virginia has the highest quality of coal found anywhere in the world and we have plenty of it (Reports of our diminishing reserve base has been wrongly placed)! We have over 52 billion tons of demonstrated mineable reserves or 350 years of production remaining at today's production levels.

The coal industry remains vitally important to our state and its economy. Together, with the states electric power industry, it accounts for nearly 60 percent of the total state business tax collections. These tax dollars translate directly into important education, government and community services and provide a reliable revenue stream for many other county, local and municipal programs.

No other state business or industry affects so many people in so many different ways! It's overall impact is staggering in terms of employment, wages, taxes and overall economic activity.

The state's industry is postured with an abundance of opportunity as the worlds thirst for low-cost, reliable energy grows on an incremental basis of nearly 2 percent annually. Thus, coal generally and West Virginia's coal particularly will continue to be a major player in the world wide energy mix on a going forward basis.

A strong energy market and high demand has created an uplifting and positive energy around the state that most of us in the business have not witnessed since the 70s—and with that optimism comes the realization that we can do so much more if we are able to capitalize on todays opportunities. West Virginia Coal will be relied upon more than ever for industrial and household energy; domestic energy independence; national strategic defense; homeland security, and today's ever popular "coal-to-liquids" and "coal conversion technologies."

And lastly, just as all miners and mine managers have come together to grieve over the tragic events of the last three weeks, they all need to be part of the solution so we may effectively prevent a similar event in the future. They all have unique experiences and qualifications to contribute!

Today, more than ever before miners, mine managers, engineers, research institutions and government officials need to become engaged to develop safer mining plans, better designed equipment and more effective ways to control our environment. Matters of safety, security and stability are shared responsibilities.

And as the industry prepares to retrain its existing workforce along with the next generation of skilled miners, the "Sago" miners will be forever remembered and serve as a daily reminder of the supreme vigilance required in the workplace!

Our membership has an abundance of safety, technical and operational expertise which has been called upon to respond to the challenges before us. We hereby extend those resources for your use and dedicate the same towards making the West Virginia Coal industry the safest in the world!

I'll close by reciting the inscription on the "the West Virginia Coal Miner" statue located on the grounds of our state capital which captures the essence and summarizes best the importance of the coal miner and coal mining to West Virginia and to the Nation . . .

"In honor and in recognition of the men and woman who have devoted a career, some a lifetime, towards providing the state, nation and world with low-cost, reliable household and industrial energy . . . Let it be said that 'Coal' is the fuel that helped build the greatest country on earth, has protected and preserved our freedom and has enhanced our quality of life. God bless the West Virginia Coal Miner."

Thank You.

Senator SPECTER. Thank you very much, Mr. Hamilton. Our next witness is Mr. Davitt McAteer, Vice President of Sponsored Programs of Wheeling Jesuit University, former Assistant Secretary of Labor, Mine Safety and Health, worked as a consultant to the former West Virginia Governor, Bob Wise, a Bachelor's Degree from Wheeling Jesuit University, and a Law Degree from West Virginia University. Thank you for coming in today, Mr. McAteer, and we look forward to your testimony.

STATEMENT OF DAVITT McATEER, VICE PRESIDENT OF SPONSORED PROGRAMS, WHEELING JESUIT UNIVERSITY, SHEPHERDSTOWN, WEST VIRGINIA

Mr. McATEER. Thank you, Chairman Specter, Senator Harkin, Senator Byrd, Senator DeWine. Thank you for inviting me to participate in this hearing on the Sago Mine disaster and on mine safety generally. It is tragic that we now must expand the scope of this inquiry to include the fatal mine fire at Massey Energies Alma Mine in Logan County, West Virginia. As you mentioned, I am presently Vice President of Wheeling Jesuit University, but previously served as the Assistant Secretary of Labor from 1993 to 2000. I have been involved in mine safety in a number of capacities since 1968.

On Friday, shortly after the Sago disaster, Governor Joe Manchin called, and asked me to conduct an independent investigation to determine the cause of that disaster, to assess the rescue and recovery operation, and to make recommendations to improve mine safety in West Virginia. In accepting that assignment, I pledged that we would take every step to identify the problems that caused the disaster that contributed to the difficulties of the mine rescue operation, as well as to the missed communication to the Command Center, and the families, and that we would fix these problems.

Today, I want to make that same pledge to the families of Don Bragg and Elvis Hatfield, the miners who lost their lives at Alma. We will hold public hearings, we will hold them on or about the 1st of March. We will have a report to the Governor on or about July 1. I understand, as Mr. Hamilton does as well, that Governor Manchin is this very morning, introducing three pieces of legislation in an effort to try and remedy problems that we have discovered and know exist as Cecil Roberts pointed out, know exist in the mine rescue area. Those are the rapid response system, more SCSR's, so you can get more time to get out, and new technologies to try and deal with them. Let me say here and now, that technologies do exist, and that is present, and available, that is off the shelf, and has been approved by MSHA.

This is the tracking system that the PED, the Personal Emergency Device that Senator Harkin mentioned, this is currently available. It is in roughly a dozen mines in this country. It can be put into all the mines in this country, and one of the proposals Governor Manchin is making is to do just that, to put this device in the mines. Had we had this device, a signal could've been sent from the surface to the miners immediately upon the learning of the danger, and to tell them to evacuate.

Second, this is the tracker system. This is a tracker that can be carried by an individual miner, and he can be located. These devices have been approved by the Mine Safety and Health Administration. I must disagree with Acting Assistant Secretary Dye. These devices have proved to be reliable. They have proved to be effective, both in this country and abroad. These devices are in mines in China. These devices are in mines in Australia. They were developed after disasters in Australia by government funded research that suggested that we need to be able to locate them. They are available to the industry, and it is my hope that the Senate and that the Administration would work to see that these devices get put into the mines immediately.

Senator HARKIN. How much do one of those cost?

Mr. MCATEER. This is the system that you put a loop—an antenna loop, a high end cost would be $100,000 for a mine the size of Sago. These devices are much less expensive, quite inexpensive; $15 to $20 is my understanding. I have literature, which I will submit for the record from the mine company.

Obviously, we're not here to endorse this one company—the technology from this one company. We are hear to say, products exists that are out there, and are not being put into the mines. Unfortunately, this mining industry, and I must say as Assistant Secretary, we did make progress. We have brought the number of deaths down, but we have not brought the number of deaths down to where they need to be. But, the industry has focused its attention on productivity, and that is their want, and that is how they make a profit, and we don't take that away from them.

But we do have to say, that the failure to focus attention on safety devices, and improvements, and put them in the mines is most unfortunate for those miners that we are all here and concerned about, and that needs to be changed. We need a safety system that will tell us at all times where miners are, and how we can get them out.

Now, let me turn to the mine rescue system. In 1995, we held a conference, a summit of all parties in the industry, labor, management, State agencies, Federal agencies, and I chaired that summit, and asked the question, are we prepared now to deal with mine emergencies? The answer is we are not adequately prepared. This is in part due to the fact, that we are a victim of our success. We do not have the horrible number of disasters that we had when Senator Harkin's father was in the mines, but we have them often enough that we need to be prepared. We identified declining numbers of people volunteering. We identified declining equipment, we identified declining interests on the part of the companies in this area.

PREPARED STATEMENT

Senator SPECTER. Mr. McAteer, your substantially over time. Could you summarize at this point, please?

Mr. MCATEER. I will sir, and I apologize.

Senator SPECTER. It's good testimony.

Mr. MCATEER. We have the bravest mine rescue teams in the world. We have not provided them with good equipment, and a good system, and that needs to be fixed.

Thank you, Senators.

[The statement follows:]

PREPARED STATEMENT OF DAVITT MCATEER

Chairman Specter, Senator Harkin, and Members of the Subcommittee, thank you for inviting me to participate in this hearing on the Sago Mine Disaster and Mine Safety.

I am presently the Vice President for Sponsored Programs of Wheeling Jesuit University. Previously I served as Assistant Secretary of Labor for Mine Safety and Health from 1993 to 2000, and I have been involved in the mine safety field in various capacities since 1968. I live in Shepherdstown, West Virginia.

As you know, the Sago Mine Disaster was really two disasters—first the crisis of the trapped miners, and then the appalling miscommunication to the waiting families, leading them to believe for nearly three hours that all but one of the miners had been saved when in fact all but one had perished. Both parts of this dual disaster must be investigated and their causes identified. And we owe it to the lost miners, and to their families, to take all necessary steps to ensure that there will never again be such a dreadful double tragedy.

Governor Joe Manchin of West Virginia called me soon after the disaster and asked if I would serve as his personal advisor and conduct an independent investigation of the cause or causes of the disaster, assess the rescue and recovery operation, and make recommendations to improve mine safety in West Virginia and across the nation.

In accepting that assignment, which Governor Manchin announced on January 9, I pledged that we would pursue every lead, follow every avenue of inquiry, and take every step necessary to identify the problems that caused the disaster and contributed to the difficulties of the rescue operation and the miscommunication to the command center and the families, and that we would fix those problems.

That is the charge of the Special Commission on the Sago Mine Disaster, and we are in the process of carrying it out. I am grateful to Governor Manchin for entrusting me with this responsibility, and I have promised to deliver a report to him by July 1.

Subsequently the leaders of the West Virginia Legislature, House Speaker Bob Kiss and Senate President Earl Ray Tomblin, appointed six legislators to work with me to help determine what went wrong at the Sago Mine and, in the Speaker's words, to decide "what we as a lawmaking body should do to minimize the chances of such a tragedy happening again." I appreciate the opportunity to work with these bipartisan colleagues, many of whom have been closely involved with coal mining.

Before going further, I want to stress that our investigation is not intended to duplicate or in any way impede the investigations under way by the Federal Mine Safety and Health Administration (MSHA), the West Virginia Office of Miners' Health, Safety and Training, and the West Virginia Board of Coal Mine Safety and Health, or by any other regulatory agency or lawmaking body with statutory responsibilities in this area. But I must also stress that our investigation, unlike the others, has a very specific constituency to whom we are accountable.

We are, of course, accountable to Governor Manchin, and through him to the people of West Virginia and the nation. But ultimately we must answer to twelve good men—Tom Anderson, Alva Bennett, Jim Bennett, Jerry Groves, George Hamner, Jr., Terry Helms, Jesse Jones, David Lewis, Martin Toler, Jr., Fred Ware, Jack Weaver, and Marshall Winans—and to the families to whom they will return now only in memory. And we are accountable as well to the sole survivor of the disaster—Randal McCloy, Jr.—and to his wife and children and all those who continue to pray for his recovery. We must answer to them.

Our responsibility, therefore, is to ask and explore all of the questions to which the miners' families must have answers, including these:

—Was the Sago Mine being operated unsafely?

—Were the responsible Federal and State agencies doing an adequate job of enforcing safety and health laws and regulations at the mine?

—What triggered the explosion on the morning of January 2?

—Could the explosion have been contained?

—Why did it take so long to get the rescue operation launched?

—Were the mine rescue teams adequately equipped and supported?

—Was the command center properly organized and directed?

—Were all communications between the command center and the rescue teams properly verified and logged?

—What caused the breakdown in communications at the mine?

—What caused the miscommunication to the miners' families?

—What were the reasons for the long delay in getting the facts to them?

—What steps must be taken to ensure that such a dual catastrophe will never occur again?

—Is the mine rescue system as effective as it should be?

—As we absorb the lessons of the Sago Mine Disaster, what additional steps must be taken to better protect miners nationwide?

To help us address these questions, I am gratified that we are receiving support from miners, mine rescue teams, miners' families, mining companies, the United Mine Workers of America, Federal and State agencies concerned with occupational safety and health, and the entire West Virginia Congressional delegation.

And, because this disaster has touched everyone in West Virginia—and indeed has affected so many people in Pennsylvania as well, and throughout the coalfields of the United States—it is also gratifying that we are being assisted by student volunteers at Wheeling Jesuit University and West Virginia Wesleyan College and by engineering and law students at West Virginia University. They are the future of our state, and it is heartening that they have volunteered to help us build a safer future for those who dig the coal that creates the energy to drive our nation's economy.

Most of our investigative work is ahead of us, and I do not want to pre-judge any part of that work or speculate about our findings and recommendations. But at the outset I do want to emphasize four points.

First, we have nothing but praise for the mine rescue teams. I have been in charge of such operations, and I know the demands that a mine rescue makes on those who willingly risk their own lives to try to save the lives of others. I'm not sure that anyone but a miner can truly understand what it's like to go underground in the full awareness of hazards seen and unseen that can instantly end your own life, and to go anyway because there are miners trapped in there who are counting on you to come after them. That is a good working definition of heroism, and we will not second-guess anyone who went to the aid of the Sago miners.

Second, we are going to get to the bottom of the Sago Mine Disaster, but we are not going to scapegoat anyone who was on the scene, including those who responded on behalf of MSHA and the West Virginia Office of Miners' Health, Safety and Training. It has been my privilege to know and work with many of those involved, and I want to say here that my strong belief, going in, is that the failures that occurred at the Sago Mine were primarily failures of the mine safety system, probably compounded by human fatigue and communications confusion, rather than failures that are somehow attributable to the personalities of those involved. These are good and competent people. I will be enormously surprised, and deeply disappointed, if our inquiry leads us to conclude otherwise about any of them. And we should all respect the fact that they are grieving the loss of their brother miners just as the miners' families are.

Third, every aspect of our investigation will be open to the miners' families and the public. We will hold a public hearing in Buckhannon, West Virginia, on or around March 1, and we may schedule additional public hearings if needed. I can see no valid reason for conducting any part of our investigation behind closed doors. We need abundant sunshine to help determine what went wrong at the Sago Mine and how to strengthen mine safety protection and enforcement.

Fourth, it is very clear that the nation has not invested as much money and energy in the cause of coal mine safety as in the pursuit of coal production and profits. That imbalance must be corrected, I believe, if we are to recover from the Sago Mine Disaster by making the mines of the United States as safe as they possibly can be made.

I do not subscribe to the fatalistic view that coal mining is so inevitably dangerous that we must be willing to accept the loss of life as part of the price of coal. Let me point out that there are productive coal mines in West Virginia and elsewhere that have operated for twenty, thirty, forty years without a disaster. Tough Federal legislation, enacted in 1969 after the Farmington No. 9 Mine Disaster and subsequently strengthened, has helped to drive down the frequency of disasters and fatal injuries in the mines. And the best mine managers today know full well that safety and productivity go hand in hand.

But legislation must be diligently and aggressively enforced to have meaning, and adequate funds must be committed to the protection of miners—both in day-to-day enforcement and in the development and implementation of technologies as advanced as those that we rely on in other fields, such as reducing the hazards of air travel, pinpointing the location of vehicles, or facilitating instantaneous, error-free communication around the globe and in outer space.

In some industries, investing in state-of-the-art technologies is considered to be just part of the cost of doing business, and it must at long last be seen that way in mining as well. The energy industry can well afford to partner with the government in ensuring that new health and safety technologies are rapidly developed and put to work in the nation's mines.

In 1995, we convened a conference on mine emergency preparedness at MSHA's National Mine Health and Safety Academy in Beckley, West Virginia. Out of that conference came many useful recommendations to improve mine rescue technology and communications, and a number of initiatives, including proposed regulatory changes, were subsequently advanced. But I share Senator Byrd's concern, which he expressed last week, that MSHA today may be "under-staffed, under-funded, and under-equipped." If that is true, it will be incumbent upon Congress to help fix the problem, so that MSHA can guide the mining industry in creating a truly modern model of mine safety.

I want to mention one initiative that deserves our immediate attention and support regardless of the findings and recommendations that may emerge from the various investigations.

It appears that precious time was lost on the morning of January 2 while company officials attempted to contact MSHA, and perhaps while MSHA organized its response. That remains to be determined, but it is clear that MSHA's emergency response system is outmoded. Mine emergency communications need to be brought into this century, in part by establishing, perhaps at MSHA's National Mine Health and Safety Academy facilities, the equivalent of the nationwide 9–1–1 emergency system that is used to rapidly dispatch emergency personnel and other first responders. We should begin immediately to develop plans for such a nationwide system and to set target dates for implementing it.

Governor Manchin is today introducing legislation to develop such a system for West Virginia. I want to commend him for his leadership in launching this initiative, and I hope that we will soon be able to expand the 9–1–1 system to protect miners nationwide.

Again, I do not want to anticipate our investigation's findings and recommendations, but I think it is already clear that certain other safety improvements deserve priority consideration. These include:

—*Technological advances in mine communications.*—If the technology can be developed to facilitate space exploration and to equip consumers with the ability to instantaneously transmit and receive information worldwide from their laptops, cellphones, and BlackBerries, there is no question that mine communications technologies can be brought out of the dark ages. The only question is whether we have the will to make the necessary commitment to research and development. It is imperative that we make that commitment and find the funds to fulfill it.

—*Improved mine safety equipment.*—The Self-Contained Self-Rescuer (SCSR), which provides an oxygen supply for at least one hour, is a technology developed more than 50 years ago. The fight to make the government require SCSRs took more than twelve years, ending with the adoption of SCSR regulations in 1981, and the sad fact is that there has been little industry support since then for efforts to improve the SCSR so that it can sustain life for longer periods of time. What is true of the SCSR is true of mine safety technology in general. In the wake of the Sago Mine Disaster an industry spokesman was quoted as saying that it was not the industry's responsibility to develop new safety equipment. "We're not in the self-rescuer manufacturing business," he reportedly said. That attitude harks back to the 19th century and has no place in the 21st. The industry may not manufacture safety equipment, but going forward it has an obligation to support accelerated research and development so that technologies available elsewhere can be transferred to the protection of miners with all deliberate speed.

—*Improved mine emergency training.*—There are fewer mine disasters today than in the past. That is the good news. The bad news is that in such a situation it is difficult for all involved in mine safety to maintain a constant state of high readiness. Mine safety officials and mine rescue teams are generally well trained, but there are few opportunities for them to test their training and proficiency in real-world situations. Among other things, we need to look into the feasibility of augmenting today's training systems and improving response times by conducting emergency rescue drills at operating mines on a surprise, unannounced basis. Doing so would, of course, involve considerable expense and inconvenience in temporarily halting production, simulating emergency conditions, and deploying mine rescue teams and support personnel. But we need to consider the trade-offs in improving emergency readiness. The paradox is that the safer we can make mining, the more we need to be fully prepared for something to go terribly wrong anyway. If, for example, the explosion at the Sago Mine turns out to have been triggered by an abnormally severe lightning strike, it will underscore the probability that we can anticipate but not entirely eliminate all hazards and must be ready—totally ready—when they suddenly put miners' lives in peril. Given the importance of those lives and our reliance on the coal they mine for our use, anything less than state-of-the-art training should not be acceptable.

I would add that there is also a need to address the issue of information control at the mine site. The Sago Mine Disaster was the first such emergency to occur in the cellphone era. While cellphones may or may not prove to be implicated in the miscommunication that occurred between the command center and the families waiting at the Sago Baptist Church, the presence of numerous cellphones at the site and the difficulty of controlling their use underscores the need to ensure that rescue and recovery information is strictly confined to the command center until it has been verified. This is not to suggest that there should be an information blackout during future rescue and recovery operations, but that only authenticated information about the progress of such operations should be released, by authority of the responsible official in charge of the command center.

Governor Manchin has repeatedly pledged that the miners who died in the Sago Mine will not have died in vain. We can fulfill that pledge only by diligently working to identify what went wrong and then by taking every necessary step to fix what went wrong by improving the safety and health protection of miners. Your decision to convene this hearing reflects your commitment to this goal and should be applauded. I look forward to sharing our investigative findings and to working together in the months ahead to make America's mines safe.

I would be glad to answer any questions and to provide any additional information that may be helpful to you. Thank you.

Senator SPECTER. Thank you very much, Mr. McAteer. Our final witness on this panel is Mr. Bruce Watzman, Vice President for Safety, Health, and Human Resources for the National Mining Association, appointed recently to the Mine Safety and Health Research Advisory Committee, has Undergraduate Degrees in Economics and Psychology, and a Postgraduate Degree in Environmental Health Management. Thank you for joining us today Mr. Watzman, and the floor is yours.

STATEMENT OF BRUCE WATZMAN, VICE PRESIDENT, SAFETY AND HEALTH, NATIONAL MINING ASSOCIATION, WASHINGTON, DC

Mr. WATZMAN. Thank you, Mr. Chairman. On behalf of the National Mining Association, its member companies in coal and mineral production, I too, once again, extend our heartfelt sympathies to the families in West Virginia. These tragedies leave no one in the industry untouched. The events have shaken the mining community, and all committed to mine safety. We join with others here today, to ensure that out of these tragedies, will emerge a stronger resolve, and greater cooperation in the pursuit of mine safety. Our expectation is that from this, and similar hearings, and from the exhaustive investigations now underway, we can do better, what we've tried to hard to do well. It's in this spirit that I appear before this subcommittee today, to offer information on where we have been, and recommendations for what we can do to advance mine safety.

As this committee considers the recommendations as a result of this hearing, I urge that we not create an atmosphere in which the parties feel the need to retreat to their respective corners of the ring, and defend themselves. Rather, let us create an atmosphere where we come to the center of the ring, stand together, and fight against the common enemy, workplace accidents.

The coal mining industry takes seriously its commitment to protect its workforce. Since the first oil embargo in 1970, the coal industry has been called upon to provide more coal to meet our energy's needs. We've answered that call while providing a safer working environment for our workforce. Since 1970, coal production has increased 83 percent, and mine fatalities have decreased 92 percent. As we look at the first 5 years of the 21st century, we see a continuation of that trend that began in early 1970's, safer coal mining. This demonstrates that safety and productivity are not competing goals, but rather complimentary goals. Working in what are inherently hostile environments, today's mining companies have proven that a well trained, experienced workforce, using state of the art equipment, can accomplish the dual goals of working safely and being productive, and yet, these accomplishments are diminished by what remains to be done, and what NMA, and its members are committed to working towards.

Following a tragedy such as these, these statistics understandably pale before the names of those lost. The events of recent weeks have strengthened our resolve to work harder, and work smarter on mine safety. This effort must begin with a close and comprehensive examination of current safety practices, and procedures. Our ability to further advance coal mine safety and health, will require an examination of the structural and technologic hurdles that must be overcome. It will require a commitment to identify, and foster the development of 21st century technology that will perform effectively and reliably in the mining environment.

Advances in technology have been integral to the safety improvements thus far, and will contribute to further advances. In pursuit of this goal, and to ensure a focused, and transparent effort, this week, the National Mining Association is announcing the formation of a National Safety Technology and Training Commission. The Commission will be drawn from safety experts for the purpose of examining technologies, emergency response, rescue procedures, and rescue conditions in our Nation's mines.

The Commission will be chaired by a recognized expert in mine safety, Dr. Larry Grayson, Chairman and Professor of Mining and Nuclear Engineering at the University of Missouri-Rolla. The Commission will report its preliminary findings by July 1, with a final report by the end of the year. We anticipate it will examine among other things, current and new promising technologies for communication, tracking miner location, rescue technology, and the methods to more reliably detect potential hazards.

This subcommittee is very aware of the need to maintain a vibrant mining research program within the National Institute for Occupational Safety and Health. The tragic events we've witnessed underscore this need. The Federal Government has an important role to play in technology development, and we urge your continued support to strengthen this vital government function. It's especially important for us to continue to work together as partners, because as Mr. Roberts said, coal is an industry with a changing face. Many of the people who joined in the 1970's, and who have built a career producing America's energy, are now retiring. We all must work together to develop programs to train and educate a new generation of employees, so that they can have a safe and productive career in an industry vital to this country's energy markets, and national interests.

PREPARED STATEMENT

Before closing Mr. Chairman, I would be remiss if I did not recognize the efforts of those who participated in the rescue activities. The efforts of these brave and often unheralded rescue team members cannot be minimized. We as an industry are fortunate to have these people as part of our mining family. We thank them for their service. Mr. Chairman, we stand in the ring with you and the members of this subcommittee ready to lead in the effort to advance coal mine safety. Thank you.

[The statement follows:]

PREPARED STATEMENT OF BRUCE WATZMAN

Good morning. My name is Bruce Watzman. I'm the Vice President for Safety and Health for the National Mining Association.

On behalf of NMA and its member companies in coal and minerals production, I extend, once again, our heartfelt sympathies to the families who lost loved ones at the Sago mine.

This tragedy leaves no one involved with the industry untouched. Anyone who has worked in a mine . . . or knows someone who does . . . feels deep sorrow.

In addition, it compels all of us in the mining community to work harder towards the one goal we all share—zero fatalities.

We join with others here today to ensure that out of this tragedy will emerge a stronger resolve and greater cooperation in pursuit of safer mines. Our expectation is that from this and similar hearings . . . and from the exhaustive official investigation now underway . . . we can do better what we've tried hard to do well.

It is in this spirit that I appear before this subcommittee today—to offer information on where we have been and recommendations for what we can do to advance mine safety.

As this committee considers the recommendations as a result of this hearing, I urge that we not create an unproductive atmosphere in which parties feel the need to retreat to their respective corners of the ring and defend themselves. Rather, let us create an atmosphere where we come to the center of the ring stand together and fight against a common enemy—workplace accidents.

INDUSTRY PERFORMANCE

The coal mining industry takes seriously its commitment to protect its workforce. Since the first oil embargo in the early 1970s, the coal industry has been called upon to provide more coal to meet our Nation's energy requirements. We have answered that call while providing a safer working environment for our workforce. Since 1970, coal production has increased by 83 percent and coal mine fatalities have decreased by 92 percent.

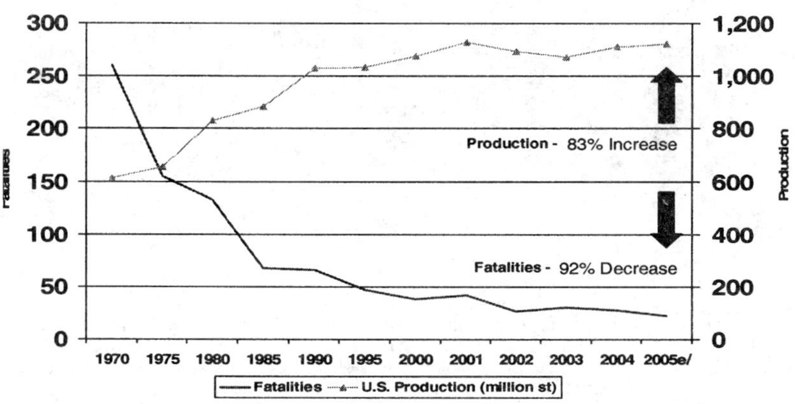

U.S. Coal Mine Safety and Production Trends

Source: Mine Safety & Health Administration (MSHA)

As we look at the first five years of the 21st century we see a continuation of the trend that began in the early 1970's—safer coal mining.

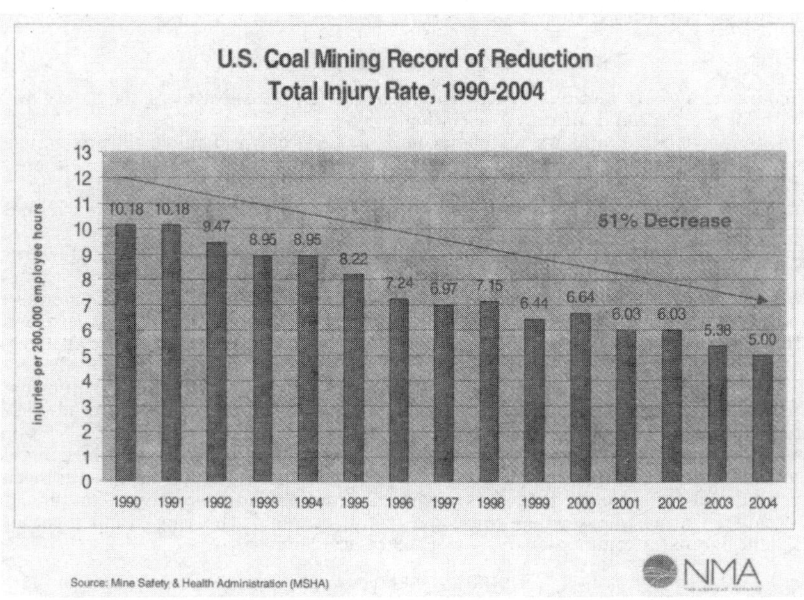

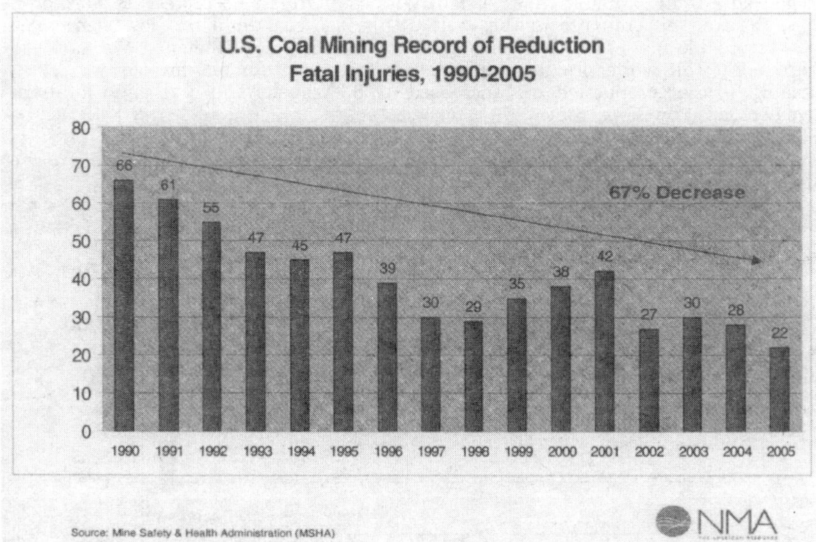

This demonstrates that safety and productivity are not competing goals, but rather complimentary goals. Working in what are inherently hostile environments, today's mining companies have proven that a well-trained, experienced workforce, using state-of-the-art equipment, can accomplish the dual goals of working safely and being productive. And yet these accomplishments are diminished by what remains to be done and what NMA and its members are committed to working towards. Following a tragedy such as this, these statistics understandably pale before the names of those lost.

MINE SAFETY AND TECHNOLOGY

The events at the Sago Mine have strengthened our resolve to work harder and work smarter at mine safety. This effort must begin with a close and comprehensive examination of current safety and rescue procedures. Concurrent with a well-trained workforce and state-of-the-art equipment, the coal industry has incorporated safety management into its business and moral ethic. Safety management comprises four functions: prevention, detection, first response and sustained response. These are not new to the coal industry; we've incorporated them for decades as sound safety practice dictates.

Our ability to further advance coal mine safety and health will require an examination of the structural and technologic hurdles that must be overcome. It will require a commitment to identify and foster the development of 21st century technology that will perform effectively and reliably in the mining environment. Technologies such as the introduction of remote control miners, integrated methane monitors on mining equipment, atmospheric monitoring systems, longwall mining systems and canopies on equipment are a few of the advances that have contributed to the industry's improved safety record. Advances in technology have been integral to our safety improvements thus far and will, we believe, contribute to further improve mine safety in the future.

MINE SAFETY COMMISSION FORMED

In pursuit of this goal and to ensure a focused and transparent effort, this week the National Mining Association is announcing the formation of a Mine Safety Technology and Training Commission. The commission will be drawn from safety experts in academia, labor, industry, and public and state agencies for the purpose of examining safety technologies, emergency response and rescues procedures and training regimes that could significantly enhance safety and rescue conditions in our nation's underground coal mines. The commission will be chaired by a recognized expert in mine safety, Dr. R. Larry Grayson, chairman and professor of mining and nuclear engineering at the University of Missouri-Rolla. Dr. Grayson will report the commission's preliminary findings to the public and mine safety authorities by July 1, with a final report by the end of this year. We anticipate the commission will examine, among other items, the current and new promising technologies for mine communication, tracking miners' locations, rescue technology and methods to more readily and reliably detect potential safety hazards.

This subcommittee is very aware of the need to maintain a vibrant mining research program within the National Institute for Occupational Safety and Health (NIOSH). The tragic events at the Sago Mine underscore this need. The Federal Government has an important role in technology development—in order to bring safer, new devices to a relatively small market for safety equipment. We urge your support to strengthen this vital government function.

In addition to government participation, our industry will continue to examine how new technology and training can be adapted to further improve mine safety performance. We must continue to use the labor- business-government model that has served us well in the past on coal mine safety.

It is especially important for us to continue to work together as partners because coal is an industry with a changing face. Many of the people who joined this industry in the 1970s, and who have built a career producing America's energy, are now retiring. We—government, industry and workers—all must work together to develop programs to train and educate a new generation of employees so that they can have a safe and productive career in an industry vital to this country's energy markets and national interests.

In conclusion, if we work together, as partners, and if we focus on improvement we will continue to advance both the cause of mine safety and the cause of energy security.

Before closing, Mr. Chairman, I would remiss if I did not recognize the efforts of those who participated in the rescue activities at the Sago Mine. The efforts of these brave and often unheralded rescue team members cannot be minimized. We as an industry are fortunate to have these brave individuals as a part of our mining family. We thank them for their service.

We stand in the ring with you ready to lead in this effort to advance coal mine safety.

Thank you, Mr. Chairman.

Senator SPECTER. Thank you very much, Mr. Watzman. Before proceeding to the 5 minute rounds for questions, I'd like to place in the record the identities of the miners who've lost their lives. Mr. Tom Anderson, Mr. Alva Bennett, Mr. James Bennett, Mr. Jerry Groves, Mr. George Hamner, Jr., Mr. Terry Helms, Mr. Jesse

Jones, Mr. Dave Lewis, Mr. Marin Toler, Jr., Mr. Fred Ware, Mr. Jackie Weaver, Mr. Marshall Winans from the Sago Mine, and also Mr. Randal McCloy, who is still hospitalized in a light coma, and the fatalities from the Alma Mine are Mr. Don Bragg and Mr. Ellery Hatfield. And on behalf of this subcommittee, and the full committee, and the Senate, and the whole Congress we express our sympathy to the families of those individuals.

Let us proceed now, to the 5 minute rounds for members, and let me begin with you Mr. Hatfield. What steps has the company taken to compensate the families of the deceased miners?

Mr. HATFIELD. That's an excellent question Senator. Let me explain what we've taken to this point—the steps we have taken, first, we immediately sent a letter to each of the family members—surviving family members, confirming that the funeral costs would be covered, and also confirming to the families, the amount of the insurance coverage they would receive, by virtue of our companies benefits program, and the life insurance.

Senator SPECTER. How much is that insurance coverage?

Mr. HATFIELD. It varies depending on the individual, from $164,000 to nearly $300,000 per miner.

Senator SPECTER. What beyond that, will the company do for the families?

Mr. HATFIELD. Beyond that, recognizing that sometimes these issues take awhile to work through the system, and the insurance companies often need information that takes awhile to provide, we immediately——

Senator SPECTER. I'm talking about the company's efforts beyond the insurance companies.

Mr. HATFIELD. We extended their salaries, they continued to get paid today just as much as they earned when their husband or brother was alive.

Senator SPECTER. How long will that last?

Mr. HATFIELD. We have extended that for 6 months, to be sure that it safely bridged the period of time when the life insurance proceeds will be delivered, and when West Virginia Worker's Compensation awards are available to the families.

Senator SPECTER. Anything beyond that?

Mr. HATFIELD. We've also extended medical care, full benefits, medical, vision, and dental through the entire COBRA period of 18 months at no cost.

Senator SPECTER. Mr. Hatfield, let the subcommittee know what, if any, additional benefits the company intends to provide here.

Mr. HATFIELD. We will certainly do so Senator.

Senator SPECTER. The issue Mr. Hamilton as you articulated, is to commit whatever resources are needed, that's a pretty broad statement, considering the necessary resources. We've seen Mr. McAteer testify about certain devices which could be very, very helpful. The thought crosses my mind, that the coal companies have been very profitable lately. Energy is in big demand. We know what has happened. Coal's an enormous resource. What about a potential for a user fee? When we wrestle with the budgetary problems, the allocations to the Department of Labor are at risk with other very, very high priorities, Pell Grants, Community Health Services, OSHA. What do you think Mr. Roberts, about the possibility of imposing a user fee to deal with your testimony about additional mine rescue teams, or what Mr. McAteer testifies about additional equipment?

The real responsibility lies with the industry, as opposed to the taxpayers generally, and if you have a user fee, it's not subject to a one percent cut, it's not subject to bargaining with Pell Grants, or scholarships, or health for children. What do you think Mr. Roberts?

Mr. ROBERTS. Let me make a couple of comments if I might, Senator, about that. One, is some of the suggestion is that have been made here today, particularly with respect to oxygen on the section. I believe that the cost of a canister is somewhere around $3,500 to $5,000 it's in that range, depending upon what type of oxygen you're talking about providing. The cost of running an extra telephone line into the mines is probably 10 cents, or 15 cents a foot, so those couple of things are minimal cost.

I think the larger cost items going to get into—and let me just make a comment if I might, about mine rescue teams, and what the law says, and somehow we've strayed away from that, and I do want to say, and he may be surprised by this, but Mr. Hatfield commented in a press conference that he intended to place onsite a mine rescue team, consisting of miners who worked at that mine, and I think that is something that the law says should be done. But over the years, we've gotten away from that, and I think he's on the right track with respect to that. The law says that there will be two teams at every mine, and within 2 hours of the mine, but we don't require that to happen.

Obviously since the—when the downturn in the economy hit in the 1980's, we got away from some of this. But we, I think, need to go back to a strict interpretation of the law. But to answer your question, our union would support any actions to fund protections for coal miners, because we should remember one thing here today, I believe Senator, when the act was passed in 1969, it said the coal miner was the most precious resource in the mining industry. So, whatever steps we have to take to protect the coal miners in this country, we should take.

Senator SPECTER. Thank you, Mr. Roberts. The red light went on in the middle of your testimony, so I would yield at this time to Senator Byrd.

Senator BYRD. Let me thank Governor Joe Manchin, his efforts have been recognized here appropriately. He has left no stone unturned in his desire to help the families of these stricken miners, and to help the industry, and the other elements that are working together to protect the coal miner. He is doing everything he possibly can.

Let me also thank my colleagues, Senator Rockefeller, and Congressman Ray Hall—Nick Ray Hall, they were at the scene, they stayed at the scene. I also want to thank the other members of the delegation for the support that they give to coal mining, coal miners, and safety of coal miners, and I'm talking about Mr. Mollohan, and I'm talking about Ms. Capito. They are supporters of every feature of the law, and appropriations that come before the Congress.

But let me say this too, I also want to thank the witnesses here today who've come. You've given splendid testimony, and you can give more in whatever you wish to put in writing for the record of this committee. You can be sure that it will be read, not just by the staff's of the Senator's, but by Senator's themselves. I particularly call to the attention of the committee that the members of the delegation would all like to have been here, but for one reason or another, some of them could not come.

My question would be to Mr. McAteer, what needs to be done to get to the problem of habitual violators? Are the penalties sufficient, or not? Do the fines need to be increased? What suggestions do you have? What needs to be done to get to the problem of habitual violators, they just pay the fine? That's not enough; they go on with the violations. That happens, and is happening. How do we get at this problem?

Mr. McATEER. Senator, thank you. Let me try to give two parts of an answer. First, as has been mentioned here by Senator Specter—Chairman Specter, that the price of coal has increased dramatically, and the price of the penalties have not increased dramatically, and as with all of us, unless it makes a difference in our monetary budgets then penalties are below what they need to be, don't have the impact that you want them to have, and I'm afraid that situation is where we have it today.

Second, with regard to the habitual violator, there is a scheme, a pattern of violations scheme within the act, which was enacted in 1969 and amended in 1977, which is an effort to get at the—at that type of violator. That has had a limited use in the past 5 years. The second part of it, it can be used in several ways. It can be used one, as a notice to the operator that that change needs to be made, and that by sending the letter, and saying we're looking at you to put you on a pattern, is one way to do it.

Another way to do it is to send an inspector with some degree of regularity to the mine, and then the mine company pays attention to the violations, because you have an inspector there. The ultimate responsibility, as everyone has testified here today though, rests with the mine operator. When you see the violations increasing, when you see the citations increasing, when you see the accidents rates increasing at the mines, then you as the mine operator must take action, and we the Federal Government and State government, have the responsibility to urge, and encourage, and use penalties to force that. But, it has to be on the part of the mine operator.

So, I would say yes to the increase in fines, and yes to the increase in use of pattern of violations. That needs to be done, and it needs to be done immediately.

Senator BYRD. Mr. Roberts, would you like to comment?

Mr. ROBERTS. I'd like to just add Senator, thank you for the question, one thing to this issue of fines, its one thing to issue fines and penalties, and quite frankly, we have a number of Federal Inspectors out there who do a good job of that. But if you do some investigating here, you're going to find something. That on appeal, those fines are reduced to almost nothing, and so there are two issues here.

One, what level should the fines be? But at the end of the day, when coal companies do spend a lot of money on attorneys, at the end of the day, if those fines get reduced to zero, the issue that I would ask you to look at, as Jim Walters resources in 2001 where 13 miners lost their lives, and the Government assessed huge fines there, but they were dropped to nothing. More or less, I think it was $3,000 dollars over a period of time, through legal work, and appeals. There's something wrong with the process when that can be allowed to happen.

Senator BYRD. A miner knows his work best. The Mine Act guarantees every miner, union or non-union, the right to leave a mine whenever he feels that his safety is threatened. How can we ensure that miners feel that they're free to exercise that right without fear of retribution? Mr. Hamilton.

Mr. HAMILTON. Thank you, Senator Byrd. My response to that question is you're absolutely correct; there are more than adequate protections under the law. The law is crystal clear, that every miner in every mine has the ability to personally withdraw himself if he thinks that his personal safety is jeopardized, or the safety of others is compromised.

I've been a coal miner, I've been around coal miners all my life. They're a very strong, courageous people. I have seen those rights exercised, and there are protections in the event that someone wants to discriminate against a miner, or any other person in our workforce for exercising those rights. There are ample adequate protections there as well. Those rights are exercised; those protections are clearly understood by the men and women who mine coal. Again quite honestly, there's an extraordinary skill, and qualification and maturity level amongst the miners in West Virginia, and these miners take great pride in their work, and if in fact there are unsafe conditions, they report those unsafe conditions. If in fact they cannot, or don't have the qualification to correct those conditions initially on their own.

Senator BYRD. Mr. Chairman, thank you. You've been very liberal with your extension of time to me. I also again, want to thank the witnesses. I think the testimony has been helpful, and will be helpful. Thank you very much.

Senator SPECTER. Thank you, Senator Byrd. Senator Harkin.

Senator HARKIN. Thank you, Mr. Chairman. Mr. Roberts, in your testimony you said something—you didn't say it, but it's in your testimony. I read it, it said just since August 2000, MSHA has records of well over 400 mine fires, ignitions, explosions, and inundations that far too easily could've developed into significant disasters and fatalities. Many other instances likely went unreported. Four hundred. Think about that, 400. Then you went on to say miners should not have to get sick, or to risk their lives just by going to work. Now, that's kind of a straight forward statement. People shouldn't have to risk their lives, or get sick just to go to work everyday, and this is why I'm trying to get to Mr. McAteer.

The one basic point to make today for the reading, Mr. Roberts, from your testimony, is that as a Nation, and as an industry, we already posses the knowledge and the ability to prevent most of the deaths that are still occurring in the coal mining industry. What is needed is a real commitment by our Government, in this case MSHA, to do better.

Senator BYRD. Amen.

Senator HARKIN. You know, that really sums it up. Now, I go to Mr. McAteer. Now you were there when I pointed out, that this one regulation on the SCSR's—self contained devices—rescue devices, you promulgated that, and I think what—1998 maybe, 1999—I'm sorry, I don't remember when. But, that was the one I asked Mr. Dye about that was done away with in December 2001, do you recall that?

Mr. MCATEER. Yes, Senator.

Senator HARKIN. Do you recall at that time, when it was withdrawn? This proposed rule came under your jurisdiction, it sat there, then it was withdrawn in December 2001. Do you recollect at that time, what you might've been thinking, or did you make any kind of statements at that time?

Mr. MCATEER. Senator, we didn't withdraw. It was done after I left.

Senator HARKIN. That's what I mean, it was done in December 2001. You were gone at that time. I'm asking you, what did you think about it at that time; you were gone at that time?

Mr. MCATEER. If I could answer it in a little broader way, I think it would be helpful to the senators. To place the self contained, self rescuer devices in the mines in the 1980's, took a lawsuit on behalf of the individual miners. The companies who were making them said they had prototypes, but not production models. The industry said we didn't have orders, so we didn't order it, and the agency wouldn't promulgate the rule.

We had the good fortune of going before Judge Higgenbotham, who was up in Philadelphia in the Third Circuit, and Judge Higgenbotham had been an engineer, and he said this is nonsense, put the devices in the mines. But, it took a lawsuit to get there. To try to enhance that—to try to enhance that device and others, it takes a tremendous amount of work. Unfortunately historically, our industry had not been the most receptive to new devices, and now when it was withdrawn, I was saddened. I was saddened because we don't have the process.

I must tell you, that NIOSH about 9 months ago, asked an institution at the university that I work with, the National Technology Transfer Center, that Senator Byrd has supported, and asked them to help us look for improvements in the SCSR. And in addition to that, NASA has agreed for us to look for new technologies at that center.

Senator HARKIN. Let me ask you about this tracking device, where was it made—do you know where it's made? I think it's made in Australia; I'll just answer my own darn question. You say it's about $20?

Mr. MCATEER. Something to that effect.

Senator HARKIN. With this $20, you could locate any miner?

Mr. MCATEER. There's a transponder that's stationed throughout the mines, and then you can locate any individual, at any time.

Senator HARKIN. You know exactly where they are.

Mr. MCATEER. That's correct.

Senator HARKIN. $20. How many miners wear these, and take these underground with them?

Mr. MCATEER. Currently?

Senator HARKIN. Yeah.

Mr. MCATEER. I don't know. It's very, very small. Perhaps none in this country.

Senator HARKIN. Do you know Cecil?

Mr. ROBERTS. If I had to give you an answer, I'd say none.

Senator HARKIN. $20, and in case a mine has two of the most important things, communication, location, and where you are, and how can we communicate? This tells you where you are. Now you've got this device here. Now this was the text messaging, is this the same device they used out in Utah, or do you know?

Mr. MCATEER. That's correct sir.

Senator HARKIN. This is what they used in Utah.

Mr. MCATEER. Exactly that device. Yes sir.

Senator HARKIN. It's got a battery, the message comes through right there, so you can talk to one another, and obviously I didn't know this, it also has a light too. Obviously, you have to have a light to be able to——

Mr. MCATEER. This is attached to the miners lamp, every miner carries one of those.

Senator HARKIN. The device is attached to the battery.

Mr. MCATEER. The advance is that part on top, which gives you ability to text message to the surface to the miner.

Senator HARKIN. Right here?

Mr. MCATEER. That's correct sir.

Senator HARKIN. So you can tell them what to do. So now you— with this, you know where they are. With this, you can tell them what to do?

Mr. MCATEER. That's right.

Senator HARKIN. Those are available. Now how many miners have this available to them?

Mr. MCATEER. We believe—we've checked with the manufacturer, and 14 mines in this country are using those.

Senator HARKIN. 14 mines are using these.

Senator BYRD. Out of how many?

Senator HARKIN. Out of how many mines?

Mr. MCATEER. 15,000.

Senator HARKIN. 14 out of 15,000?

Mr. MCATEER. That's with metal and nonmetal, if you can't metal, nonmetal at the locations, as well as coal locations.

Senator HARKIN. You said it would cost about $100,000 a mine to hook that up, that sounded high to me, because I know about the antenna loops that go with it, and that seemed to be kind of high, just putting in an antenna.

Mr. MCATEER. That's the manufacturer—gave us that amount, and said that is the high end.

Senator HARKIN. That sounds like a high end to me.

Mr. MCATEER. It's not that much, but I'm suggesting to you that that would be the size of the Sago Mine, $100,000.

Senator HARKIN. That would be the loop, and buying these devices for every miner?

Mr. MCATEER. Yes sir, I'm sorry. That's the entire installment. That's what they told us.

Senator HARKIN. It seems like a small price to pay per mine.

Mr. MCATEER. Yes it is.

Senator HARKIN. How many miners at Sago? How many miners?

Mr. MCATEER. 150.

Mr. HATFIELD. Approximately 150.

Senator HARKIN. So, 150. You're talking about less than a $1,000, $750 per miner to be able to have communications, $20 to tell us where they are. It seems to me, this is the investment we ought to be making. I can't believe in this day and age with the technology that we have, even with the oxygen breathing devices, we know we can develop oxygen breathing devices that are longer than an hour. Much longer than an hour. Why aren't we doing it?

Mr. Roberts said a miner shouldn't have to get sick, or fear death just to go to work everyday. This is just—I don't know Mr. Chairman, it just seems to me that you hate to regulate everything, but doggone it, if they won't do it, we got to tell them to do it. It seems to me, this is something that every miner ought to have available, period. Thank you, Mr. Chairman.

Senator SPECTER. Thank you, Senator Harkin. Senator DeWine.

Senator DEWINE. Thank you, Mr. Chairman. Mr. McAteer, you did describe in your written testimony how the technology for this self contained, self rescuers, which provides oxygen supply for an hour, was developed more than 50 years ago.

Mr. MCATEER. That's correct.

Senator DEWINE. Could you tell us a little bit, where is that technology now, and where should it be going, and what's the potential here, and maybe Mr. Roberts, you could comment as well. You talked a little bit about where it ought to be, and I'd like for you to comment—one of you, then the other, then I have a couple of other questions if you can do it real quick.

Mr. MCATEER. Yes sir. There are two types of technology, bottle oxygen and potassium super oxide—a device that you breathe into, and it gives you back oxygen. Those two technologies are advanced. We have them, we are examining under this NIOSH grant, the industry where those are, and what advances have been made. We think there has not been a large expenditure of funds on new technologies. We think that there are some advances that we can bring in, and we can add numbers of ours. I can't tell you what the outside limit of that would be, but we believe that we could give you longer periods of time, for less weight, which is one of the problems. The advances that we are putting in place in West Virginia today are to add devices, so that you can carry two or three, and get from one location to another.

Senator DEWINE. Mr. Roberts, do you want to comment about that, and also while I've got you, let me just say you're testimony was very concise, very good, very helpful. If you had to prioritize, what would you prioritize? Give me one, two, three, four.

Mr. ROBERTS. The most immediate thing we could do, I believe, to give the miners the best opportunity to survive, and I think that is where we have to start, is to immediately put additional oxygen devices underground. That is point one.

Point number two, and I don't want to speculate on what could've happened, but I think I heard Mr. Hatfield say something similar to this, I think experts looking at this will tell you if there had been an ability to talk to these miners, I probably would've told them, you're best opportunity for survival is to leave, and we could've even sent mine rescue teams toward them, as they walked outside. Perhaps, we would've had more miracles here today.

So, I would start with the oxygen, I would go to the communications, and we have to Senator, begin right this minute to deal with this mine rescue team, because Mr. McAteer talked about 1995 when he brought all of the people together in this industry, I would submit to him, and to you, and to everyone else the problem is worse right now, than it was in 1995 when he knew then that we had a problem. The problem is much, much worse. We have difficulty finding enough mine rescue teams in these disasters to deal with these problems.

Senator DEWINE. Mr. Hatfield, let me ask you, did you have any difficulty in contacting the Federal Government for rescue assistance on the day of the accident, and how would you preliminarily assess their involvement in the rescue and recovery efforts?

Mr. HATFIELD. Senator, we did have some initial difficulty, because of the nature of the holiday season, in that our first calls to the various offices were going into a voicemail and the like, but there were messages on those voicemails that gave us alternate numbers to call, and so the process took a little longer than it might otherwise have. But, we were on the phone with the State regulators by I think around 7:50, and we had contacted MSHA by about 8:28.

Senator DEWINE. You lost a little time.

Mr. HATFIELD. We lost a little time, but they were responsive through cell phones and home numbers, because these are mining professionals, they want to be available. They always give contact numbers; it just took us a little while longer to get there.

Senator DEWINE. I wanted to make sure we understood what the situation was there. Mr. McAteer, where's the responsibility now? Where do we go in regard to the different technology that you've talked about? How does this get done? We've talked about several different technologies, how do we move forward now?

Mr. MCATEER. Senator DeWine, that's the problem. You put your finger on a problem. We have not spent the money, neither the industry, nor the Federal Government has spent the dollars necessary to advance the technology, and we need to do that. I think both the industry and the government, at a Federal and State level need to be doing that to see that that happens.

This Australian device I mentioned grew out of research that was done after mine disasters in Australia, and that was paid for by the Australian Government. But I think we have this problem here, where we have a gap, and we don't have the kind of enhancements in technology that we should be doing, and it's not on a continuing basis. We might do it for one thing like the SCSR, but then there's never—we don't advance it, and we don't bring in technologies from NASA, the Department of Defense into the mining community.

Robots for example, that can go on the moon, we've tried to introduce into mine rescue systems, because it's better to send a robot down, then your not so much concerned about the fires and the explosions. But we have only used that on one or two occasions, and we haven't made the enhancements, and it's expensive, but we haven't made the advancements that we need to be making in that area.

Senator DEWINE. I thank you. Thank you, Mr. Chairman.

Senator SPECTER. Well, thank you very much gentlemen.

Mr. HATFIELD. Mr. Chairman, Ben Hatfield. Could I ask the indulgence of the committee to make a brief comment on the technology, because we certainly share that interest with you.

Senator SPECTER. Go ahead Mr. Hatfield.

Mr. HATFIELD. The PED system Mr. McAteer has referred to, has certainly holds much promise. Indeed, we have that system installed at our Illinois Mine, the Viper Mine, and have seen some success with it. The couple of limitations that certainly we want to focus on in moving forward with this effort is that it's a one way communication. There are reliability issues. But first and foremost, with respect to the Sago incident, we have not found the technology yet, that will allow us to communicate through 300 feet of rock. When you don't have a line on sight, unless there's a cable or an antenna that goes underground, and there's the challenge. Because in an explosion—the nature of the Sago explosion, it obliterates everything in its path. Antennas and cables were immediately gone in an instant, and that is the challenge for the technology advancement efforts here to find ways, do not rely on things that are going to be damaged in an explosion, and that's something we certainly want you to understand. But our Viper Mine in Illinois is making use of this system, and I think it does hold significant promise.

I would also like to clarify, and apologize for my earlier answer to you, not being as thorough as it should have been. One other step that International Coal Group has taken with respect to helping the Sago families is the establishment of the Sago Mine Fund. We made our own initial contribution of $2 million, and that fund, through the generosity of many others across the country, has already $20 million, and that is exclusively dedicated for the benefit of the Sago victims, and we expect that fund to continue growing. Thank you.

A letter from Senator Durbin asking about the safety of the Viper Coal mine, with my response follows:

UNITED STATES SENATE,
Washington, DC, January 8, 2006.

BENNETT K. HATFIELD, *President and CEO, International Coal Group, 2000 Ashland Drive Ashland, KY.*

DEAR MR. HATFIELD: I am writing to you to determine the safety condition of the Viper Coal Mine in Williamsville, Illinois that is owned by your International Coal Group (ICG). It has come to my attention that the Viper Mine is one of 12 coal mines that the ICG owns and operates, including the Sago Coal Mine in West Virginia where a mine accident recently killed 12 employees.

I am deeply saddened by the tragic loss of life in West Virginia. Like many others, I also am disturbed by the egregious number of citations issued to the Sago Coal Mine in the year prior to the mine accident for "serious and significant" violations of Federal safety laws. Coal miners accept that their occupation is risky. But they have a right to expect that their employer will put safety ahead of other considerations, and the ICG has a moral obligation to ensure that unsafe practices are scrupulously identified, promptly corrected, and proactively avoided. The ICUs safety record at the Sago Mine raises questions about the company's commitment to safety. Moreover, Federal regulators appear to be unable to prevent companies from amassing large quantities of safety violations.

According to Mine Safety and Health Administration records, the Viper Mine in Illinois was cited for 42 "serious and significant" violations out of 124 total citations in 2005, which was the f rst full calendar year in which the ICG owned the mine. It is my hope that the Sago disaster was an isolated incident and does not reflect a pattern of mismanagement by the ICG and unsafe conditions at the Viper Mine.

At a time like this, the burden is on the ICG and our nation's regulatory authorities to ensure that the ICG is properly and safely managing its mine in Illinois, Please provide to me and the people whose lives are affected by operations at the Viper Mine a safety status report on the Viper Mine. In addition, please include the safety record of the mine since the ICG purchased the operation and the extent of any steps the ICG has taken to address past safety violations and what the ICG plans to do to prevent them in the future.

Thank you for your time and attention to this matter. I look forward to hearing from you within the next month.

Sincerely,

RICHARD J. DURBIN,
U.S. Senator.

INTERNATIONAL COAL GROUP, INC.,
Ashland, KY, January 16, 2006.

Hon. SENATOR RICHARD J. DURBIN,
U.S. Senate, 332 Dirksen Senate Office Building, Washington, D.C. 20510–1304.

Re: Your letter of January 8, 2006 regarding ICG Illinois, LLC's Viper Mine

DEAR SENATOR DURBIN: Thank you for your concern for the safety of the employees at the ICG Illinois, LLC ("ICG Illinois") Viper Mine. Although your interest in mine safety at Viper seems to have been prompted by the Sago mine disaster at an affiliated subsidiary of ICG Illinois' parent company, International Coal Group Inc. (ICG), I am pleased to have an opportunity to introduce you to ICG's and ICG Illinois' commitment to workplace safety.

Misapprehension of the facts about ICG subsidiary safety records is understandable given the recent media-induced rush to judgment about the safety record at the Sago Mine, with no meaningful review of that record, or discussion of safety at ICG subsidiaries as a whole. The ICG family continues to grieve concerning the devastating Sago tragedy. Our thoughts and prayers remain with the families of the miners who lost their lives and to the single survivor, Randal McCloy, Jr., and his family.

To give you a more balanced appreciation of ICG's overall safety record and commitment, I am attaching a one page summary of the numerous safety nominations and awards ICG subsidiaries received in 2004 and 2005. (See Attachment A). I would especially like to bring your attention to the Sentinels of Safety Award won by ICG Knott County, LLC (ICG Knott County) in 2004. This is the premier mine safety award in the nation. The Sentinels of Safety Award is an incredibly prestigious safety award with an 85 year lineage, originating under the presidency of Herbert Hoover. Simply put, ICG Knott County's Clean Energy Underground Mine was the safest underground mine in the large mine category in the entire U.S. during 2004. The Clean Energy Mine had 128,525 manhours of operation without a single loss time accident in 2004. Needless to say, ICG is extremely proud of Clean Energy's safety accomplishments, and its safety performance is a model ICG strives to emulate at all of its subsidiary operations.

ICG Illinois purchased the Viper Mine from the bankrupt and now defunct Horizon Natural Resources Company on September 30, 2004. Over the past fifteen months ICG Illinois has invested over eleven million dollars ($11,000,000) of capital in the Viper Mine, including upgrades to the mine's carbon monoxide (CO) monitoring system, the underground communications system, a new rock-dust system, addition of a second mine master supply system for both production and fire fighting, and upgrades to other underground equipment.

The Viper Mine is well known for its immediate response to both State and Federal regulatory directives during inspections. In 2005 MSHA conducted 257 inspection shifts at the Viper Mine and the State of Illinois over 40 inspection shifts. The MSHA district inspectors and State inspectors are highly regarded for their expertise in the mining industry. The Viper mine enjoys a positive, constructive relationship with both agencies.

Viper's employees are well trained, well educated coal miners who understand safety. Each employee is aware that, if necessary, they are authorized to shut down all or any part of the mine if an unsafe condition is identified. Employees may also address concerns anonymously. An employee can also contact the Illinois Department of Natural Resources (IDNR) directly or call MSHA's hotline. International Coal Group also has provided every employee a card with a hotline to call to report safety concerns, environmental issues, harassment, discrimination or other concerns.

In fact, Viper is proud to host a fully certified "Mine Rescue Team". As you may know, the Viper Mine Rescue Team was one of the teams that was dispatched to the Sago site and, while putting their own lives at risk, proudly took part in the attempt to rescue the trapped miners.

Regarding your concerns about safety citations, please refer to the tabulated data below. This list consists of safety data for nine of the older, larger Illinois underground coal mines for 2005. As you can clearly see, the Viper mine ranks at the top of its peer group.

Mine	Employees	Total citations	Number of citations significant and substantial
Viper	243	124	42
Mine A	341	138	52
Mine B	229	167	81
Mine C	151	184	50
Mine D	277	173	42
Mine E	221	227	105
Mine F	383	397	112
Mine G	253	777	277
Mine H	867	1,456	389

A serious infraction that would jeopardize miner's lives or could cause a disaster results in a closure order by State or Federal authorities which remains in place until the infraction is corrected. The larger a mine becomes, the more area there is to inspect for safe working conditions, with size accordingly increasing the number of inspection shifts per year. For example the Viper Mine has four operating sections and over ten miles of conveyor belts transporting the coal out of the mine, thus resulting in a rather high number of inspection shifts.

In December of last year, I presented my vision of ICG to our operating subsidiary management. A critical component to this vision includes being the employer of choice by providing employees a safe, secure, competitive paying job where they are treated with respect.

I extend an open invitation to you to visit the ICG Illinois Viper Mine to see first hand ICG's and ICG Illinois' commitment to the safety of its employees. I am confident that your doing so will convince you, even more than what I have shared with you in this letter, that the attacks upon ICG's safety commitment are unwarranted.

ICG Illinois, like all other ICG subsidiaries, will continue to operate the Viper Mine with a meaningful "Safety First" philosophy and practice. It is our principal priority for protecting our most valuable resource—our employees.

Thank you for the opportunity to share with you ICG's safety accomplishments and its safety commitment. Should you have any further questions, please do not hesitate to contact me.

Very truly yours,

BENNETT K. HATFIELD,
President and CEO.

SAFETY AWARDS

Year	Subsidiary Unit	Award Description
2004	ICG Addcar Systems, LLC ..	Bridger Highwall Miner Operation—Perfect Safety Record for 59,759 Man-hours with no LTA's in 2004, presented by the State of Wyoming.
2004	ICG Addcar Systems, LLC ..	Bridger Highwall Miner Operation—Nominated for a Sentinels of Safety Award for having no LTA's in 2004.
2004	ICG Eastern, LLC	Birch River Preparation Plant—Safety Achievement for 2,257 Days Without an LTA, presented by the State of West Virginia.
2004	ICG Hazard, LLC	County Line Mine—Nominated for a Sentinels of Safety Award for having no LTA's in 2004.
2004	ICG Hazard, LLC	Rowdy Gap Mine—Nominated for a Sentinels of Safety Award for having no LTA's in 2004.
2004	ICG Hazard, LLC	Thunder Ridge Surface Mine—Surface Safety Award, presented by the Kentucky Coal Association.
2004	ICG Hazard, LLC	Tip Top Mine—Nominated for a Sentinels of Safety Award for having no LTA's in 2004.

SAFETY AWARDS—Continued

Year	Subsidiary Unit	Award Description
2004	ICG Knott County, LLC	Clean Energy Mine received the Kentucky Coal Association's Surface Safety Award for the best underground mine in Hazard District.
2004	ICG Knott County, LLC	Clean Energy Underground Mine—The 2004 Sentinels of Safety Award for the Safest Underground Mine in the United States, presented by the National Mining Association and MSHA.
2004	ICG Knott County, LLC	Classic Underground Mine—Best Safety Record in Martin County, presented by the Kentucky Office of Mine Safety.
2004	Sentinel	Sentinel Preparation Plant—Best Safety Record in 2004, presented by West Virginia Holmes Safety Association.
2005	Vindex Energy	Vindex Douglas Surface Operations—Safest Mine in the State of Maryland for 2005, presented by the Maryland Coal Association.

Senator SPECTER. Well thank you, Mr. Hatfield. I'm a little disappointed and a little surprised to note that Mr. Dye and Mr. McKinney have departed the premises, so we're going to ask Mr. Friend and Mr. Clair to supplement for the record on a number of questions. We note, that consistent with what Mr. Roberts testified to, when there was a fine imposed on the Alabama incident where 13 miners were killed in December of 2001, a fine was imposed of $435,000, and that an administrative law judge reduced it to $3,000, and a number of us wrote a letter to the Secretary of Labor, dated November 22, 2005, protesting that, and asking that an appeal be taken, and without objection, that will be made a part of the record.

[The letter follows:]

UNITED STATES SENATE,
COMMITTEE ON HEALTH, EDUCATION, LABOR, AND PENSIONS,
Washington, DC, November 22, 2005.

Hon. ELAINE CHAO, *Secretary, U.S. Department of Labor, Washington, DC.*

DEAR SECRETARY CHAO: We are writing to express our deep concern about the recent decision of the Mine Safety and Health Review Commission to drastically reduce the fines sought by the department against Jim Walter Resources, Inc., in the case involving the deaths of thirteen miners at Blue Creek No. 5 Mine in Brookwood, Alabama.

We urge you to appeal that decision. The men and women working in America's mines do so in an extremely hazardous environment and they are entitled to vigorous enforcement of the Federal workplace safety laws. In the tragic explosion at Blue Creek, our Nation's deadliest mine disaster in over 20 years, the Mine Safety and Health Administration sought fines of $435,000. Fines of this level were clearly justified due to Walter Resources' long history of safety violations which led to perilous, and ultimately fatal, workplace conditions.

Incredibly, however, on November 1, an Administrative Law Judge reduced these fines from $435,000 to a mere $3,000—a decision that harms workers and erodes MSHA's authority to protect the health and safety of America's miners.

President Bush has clearly recognized the importance of fines in deterring safety violations and enforcing worker safety laws with his budget request for fiscal year 2005 to increase the monetary penalties imposed by MSHA. We urge you to appeal the unfortunate decision in the Blue Creek Mine case, and we look forward to hearing from you about your action.

Sincerely,

ARLEN SPECTER,
JOHN D. ROCKEFELLER IV,
EDWARD M. KENNEDY, AND
ROBERT C. BYRD.

Senator SPECTER. The subcommittee would like to know what has happened on that. Other questions, which we had intended to press on Mr. Dye, where reportedly 16 failures over the last 2 years at Sago, were seriousness enough to be designated as "unwarrantable failures" which has been defined as "aggravated conduct constituting more than ordinary negligence", and we would like to have a specification for what they were, and whether there was any cause of the effect. We've also noted the reduction in the number of criminal prosecutions recommended from 38 in the year 2000, to 12 in the year 2004. We would like to have some specification as to what the recommendations there were on criminal prosecutions.

One other question that we have for Mr. Dye, we note that he's the Acting Assistant Secretary of Labor, and of course the Assistant Secretary is a confirmable position, confirmable by the Senate, and I'd be interested to know if Mr. Dye has any aspirations to move from Acting Secretary to Secretary of the Department. We will be pursuing these issues further. There have been a number of suggestions made as to possible legislative action. The act, we're informed, requires the company to be present at the time that witnesses are interviewed. My own instinct is that they ought to be interviewed privately, to have the maximum freedom in answering questions.

May the record show a couple of nods from the panel, Mr. McAteer and Mr. Roberts. We're going to explore the issue of increased penalties.

Senator BYRD. Were the nods, Mr. Chairman, affirmative nods, or negative?

Senator SPECTER. Affirmative nods. Thank you, Senator Byrd. Do you think I would've commented if they'd been negative nods?

You didn't nod Mr. Hatfield, what do you think about having a requirement that the company representative be present at the time witnesses are questioned? Isn't that a little intimidating?

Mr. HATFIELD. Well I don't think so, because again, we've been very open with our people about what we know about the investigation, and have encouraged them to speak out.

Senator SPECTER. There's nothing to stop the company from interviewing them. There's nothing to stop anybody from interviewing anybody. Nobody has to answer a question, unless they're under a subpoena. You don't even have to answer a question if the subcommittee asks you to stay, you could just walk right out.

Anybody can ask anybody about everything, but nobody has to answer, unless there is a subpoena, because with subpoena you don't have to answer either if you take the privilege against self incrimination, plus your granted immunity. But why should an employee have to have a company official present? The possibility of retaliation, I know your company wouldn't, but some company might. What do you think?

Mr. HATFIELD. I can't speak with any knowledge as to what brought the rules to the position they're in, but I can speak on behalf of my company, that we certainly have no fear of leaving the interviews directly between the government representatives, and our people. We've encouraged them to be very open and honest with everything we know, and we certainly willing to absent ourselves from all the interviews, so long as the people interviewing our people, are truly the Mine Safety and Health Administration group, and the West Virginia Department of Miner Health and Safety.

Senator SPECTER. A second item, we're going to look at increased penalties. I'll give you a chance to reply to that Mr. Hatfield if you want to. Do you think it would be time to increase the penalties, perhaps commensurate with the increase coal prices?

Mr. HATFIELD. I think it's certainly a reasonable area of review. We have no objection whatsoever to review of the penalties, because I suspect they've not been visited in some time. But I would echo a comment that made by Mr. Dye earlier, that real penalty is the threat of shutting down a production section because that gets far beyond the fine level.

Senator SPECTER. I was just about to come to that. Do you endorse that?

Mr. HATFIELD. I believe that is an appropriate tool for enforcers and regulators to use, and we certainly saw it used at the Sago Mine, and other mines. Yes sir.

Senator SPECTER. How about the idea of user fees on the mines to pay for all of these investigations, and rescue teams, and all of the other safety procedures which may be an order? They really go to the benefit of the mine companies that earn the profit. What do you think about a user fee?

Mr. HATFIELD. I think it's certainly an area that is worth looking into. I don't know many of the facts, or the premises, but certainly it's an area that's fair for discussion, and advancement.

Senator SPECTER. How about Senator Harkin's photo op with this orange mechanism, do you think they ought to be required?

Mr. HATFIELD. I think that's remarkable opportunity.

Senator SPECTER. I take that to be a yes. But one other thing we're going to do, is ask the GAO to undertake a follow-up study. At the request of the subcommittee, and others, we made that request of GAO in November 2002, and they filed a report in September 2003, and we're going to request a follow-up study of the September 2003 mine safety report focusing on underground coal mines. Anybody on the panel, do the Senators have any further questions they'd like to ask at this time?

Senator BYRD. Yes, Mr. Chairman, if you will. Mine safety is a moral imperative. These miners are not to be considered expendable. What do we need to do right now to ensure a tough enforcement? Now there's been some talk about penalties, what do you think of assessing penalties as a percentage of the profits? What do you think of companies paying fines up front, so that penalties cannot be reduced by negotiation? Does anybody want to volunteer to respond? Mr. Roberts.

Mr. ROBERTS. I think we have a long history of support for the toughest enforcement possible. It's our view that complying with the law does not interfere with profits whatsoever. If you're running a safe mine, you have nothing to fear. From increased fines, we also would encourage that these fines could not be dramatically reduced on appeal, as has happened time, and time, and time again. I think that point that was made by Senator Specter, with respect to shutting down mines, I think that's the ultimate weapon in the hands of MSHA Inspectors, and I think that Congress should mandate that the law that was passed 1969, and was amended in 1977, be complied with.

Senator BYRD. Anybody else want to comment?

Mr. MCATEER. If I might Senator. We have a good law on the books. It is a law that has an enforcement aspect of it that is quite strong. In recent times, that enforcement aspect has not been emphasized as much as has cooperation. Now there's nothing wrong with cooperation, and I'm in favor of it, and I think it should be, but to do that and to diminish enforcement while cooperation takes precedence, I think is an unfortunate turn of events, and is not what this Congress, when it passed the law in 1969, and amended in 1977, had in mind.

Senator BYRD. Mr. Chairman, I simply want to say again, I think this has been helpful. Where is the point that we should attack? Should it be the Federal agency that oversees safety? Is that where the problem is? I think there ought to be stricter enforcement. There ought to be assurance that the agency will do its best to do its duty.

It seems to me that there is a kind of a mental attitude here, and mind set that yes, it is dangerous, mining is dangerous, but these violations will pay the fines, but maybe they can be negotiated downward, we'll just keep on doing it. I think there ought to be tough enforcement with this idea in the background that we're going to close you down. Why shouldn't this mine have been closed down? Does anyone want to respond?

Mr. ROBERTS. It is the question, should the Sago Mine—I think in trying to be as fair as I can be, and as objective as I can be, I think it would be until a complete investigation is completed, perhaps we don't know the answer to that, but I do support Senator, as I have just previously said, that if there is a pattern here, which is part of the law that was alluded to previously, I think that is the ultimate penalty, and it should be carried forward.

Senator BYRD. Mr. McAteer, did you want to respond to Mr. Hatfield's remarks regarding the reliability of safety devices?

Mr. MCATEER. Yes, I do Senator, and I have to take exception to his remarks. These devices have proved to be reliable. They are not perfect. It would be, in a perfect world, better to have a device that could communicate both ways. This device communicates one way, it is a step. The way we improve technology in this country, is not—the perfect is the enemy of the good. We have to put in what we have, and get them there, and then make that very next step, and unfortunately Mr. Hatfield's comments are suggestive of what we've worked with in the past. Some in the industry have had an attitude of suggesting that they won't take that step. They'll wait until the perfect is in place, and that day never comes. That's the unfortunate part of our situation at the present time.

Senator BYRD. Mr. Hatfield, did you want to take 30 seconds for any response to what has been said?

Mr. HATFIELD. I certainly agree with Mr. McAteer's point that we can't wait until the solution is perfect to implement it. I don't think any of us around this table believe we can wait for perfection. We've got to make our coal miners safer today, and that's got to be the utmost priority. But, I believe we continue to have to deal with the challenge of how do you keep an explosion from blasting away the antennas, and the cables, and everything else that we use to support these systems, and still maintain that line of communication through 300 feet of rock. I'm only pointing out there are technical issues that we need to move forward with, and that's the first one to tackle, and we certainly want to tackle it.

Senator BYRD. Thank you.

Mr. MCATEER. If I could on that question, there is no question that part of the mine is destroyed in an explosion, and at the time of an explosion, part of the mine is not destroyed at the time of the explosion. The concept behind the internet is that we have multiple nodes, so if part of it goes down, all of it doesn't go down. That's the concept that we need to adopt in the mines, multiple redundant systems that can be overlaid. In the case of the transponder, and this device here, the mine company—the Australian mine company that puts this together has suggested, that if you lose communication they can take a loop antenna, and put it on the surface, and then you can reignite the ability to communicate underground from the surface, and they've done that in other countries. Redundancy.

Senator SPECTER. Thank you very much, Senator Byrd. Senator Harkin, did you have anything else?

Senator HARKIN. No. Just when you were speaking Mr. Hatfield, I thought about redundancy. That is what you do, you build in redundancy, and there may be some technologies I'm not aware of. I'm not aware of this surface device you're talking about, but I'm sure there are new technologies out there of which I'm totally unaware of that help.

Mr. HATFIELD. I thoroughly agree. Redundancy can be part of the solution.

Senator SPECTER. Redundancy is something Senator's know a lot about.

Well, thank you very much. Mr. Hatfield, Mr. McAteer, Mr. Hamilton, Mr. Roberts, Mr. Watzman, thank you very much for coming in today, and we will be pursuing the matter, I know all of you will be too.

ADDITIONAL PREPARED STATEMENT

The subcommittee has have received a prepared statement from the National Stone Sand & Gravel Association that will be included in the record at this point.

[The statement follows:]

PREPARED STATEMENT OF CHARLES E. HAWKINS, III, CAE EXECUTIVE VICE
PRESIDENT AND COO, NATIONAL STONE SAND & GRAVEL ASSOCIATION

Mr. Chairman, the National Stone Sand & Gravel Association (NSSGA) appreciates the opportunity to submit a statement for the record of this oversight hearing on the Mine Safety and Health Administration (MSHA) and its regulatory and enforcement programs.

Based near the nation's capital, NSSGA is the world's largest mining association by product volume. Its member companies represent more than 90 percent of the crushed stone and 70 percent of the sand and gravel produced annually in the United States and approximately 115,000 working men and women in the aggregates industry. Sale of natural aggregates (crushed stone, sand and gravel) generates nearly $38 billion annually for the U.S. economy. The estimated output of aggregates produced in the first half of 2005 was 1.3 billion metric tons, a four percent increase over the same period in 2004 (2.85 b MT). According to the U.S. Geological Survey, the significant increases in aggregates production were due to the increase in construction activity, which has risen every year for the past decade. Construction spending amounted to $617.9 billion during the first half of 2005, a nine percent increase over the same period in 2004.

Aggregates are used in nearly all residential, commercial and industrial building construction and in most public works projects, such as roads, highways, bridges, railroad beds, dams, airports, water and sewage treatment plants and tunnels. While the American public pays little attention to these raw natural materials, they go into the manufacture of asphalt, concrete, glass, paper, paint, pharmaceuticals, cosmetics, chewing gum, household cleansers, and many other consumer goods.

The events in the Sago Mine and Aracoma Coal Alma No. 1 Mine disasters are tragic and the loss of even one life, let alone 14 lives, is devastating. Nevertheless, the safety record of the mining industry, and the aggregates industry in particular, has improved due to a heightened level of effort invested by the industry to sustain an improved performance. The improvement in the aggregates industry safety record is attributable to a combination of more effective safety and health programs developed and implemented by the industry over the past decade in response to increased Mine Safety and Health Administration regulation of the mining industry and heightened enforcement.

The first priority for the aggregates industry is and will continue to be the safety and health of its miners. The industry recognizes that its employees are its most valuable asset, an asset that must be protected for the well being of the industry now and in the future. As the workforce ages, it has become increasingly difficult to recruit new miners to the industry. Maintaining an excellent safety record through the implementation of effective safety and health programs is considered a critical element for attracting and keeping a highly skilled workforce. Members of the National Stone, Sand & Gravel Association have developed and agreed to a set of guiding principles to assist member companies in their efforts to understand the importance of safety to their individual organizations as well as to the industry as a whole. In addition, a safety pledge was developed in 2002 incorporating the safety guiding principles. More than 90 percent of the operations of NSSGA member companies are now covered by this pledge, signifying the importance of safety and a commitment toward ensuring the safety and health of all their employees.

Recent news articles have ascribed some of the responsibility for the Sago incident to the cooperative alliances MSHA has signed with the industries it regulates, implying an inappropriately close relationship. We would argue the opposite. The NSSGA and MSHA formalized the first such alliance in 2002, setting forth a cooperative agreement to develop programs and tools for the improvement of safety and health in the aggregates industry. The resulting reduced incidence rates speak for themselves.

It should also be noted that MSHA has similar alliances with labor organizations, including the International Association of Bridge, Structural, Ornamental and Reinforcing Iron Workers and the International Union of Operating Engineers. Important alliances also exist with the National Safety Council and the American Society of Safety Engineers. Through these alliances, MSHA has been able to enhance its mission of protecting worker safety and health.

Another collaborative effort resulted in the MSHA Part 46 "Training and Retraining of Miners" regulation in 2000. This excellent regulation ensures every miner knows and understands how to perform their job safely by covering the important safety and health information prior to starting work and annually thereafter. This regulation was developed collaboratively, with input from both labor and industry groups, guaranteeing support of the rule by all involved stakeholders and assuring their commitment to the ultimate goal of injury reduction. The Coalition for Effective Miner Training included many industry groups working in a joint industry/labor arrangement in conjunction with MSHA to develop an effective standard for the aggregates industry. The Part 46 regulation resulted from this effort.

In another example, the NSSGA and MSHA developed a cooperative workplace-based sampling training program of noise and dust monitoring workshops. A partnership agreement was signed and the training workshop program launched on December 1, 1997. These workshops have been given to industry representatives using training specialists from the Mine Safety Academy every year since 1997. These workshops have won two awards from Innovations in American Government for this joint venture aimed ay reducing hearing loss and silicosis through a program of recognition, evaluation and control of workplace hazards.

The NSSGA/MSHA alliance was further enhanced by an ad hoc coalition consisting of the U.S. aggregates industry (NSSGA and MSHA) and the quarrying industry (Health & Safety Executive and the Quarry Products Association) in the United Kingdom (England). This informal alliance was developed to share best practices between the countries in a similar industry.

Based on the sharing of information about successful programs in the UK, the NSSGA/MSHA Alliance has moved forward with joint efforts to implement programs that will further improve the safety and health of U.S. aggregates miners. The alliance first assembled a Data Mining Task Force to review the incident data (not fatalities) with the hope of elucidating specific areas where efforts could be targeted to reduce injuries. It is this focus on incidents, rather than the focus on fatalities, that offers the best chance of improving the safety performance and at the same time reducing fatalities.

Simultaneously, the alliance began working on a model safety and health program to take the best of industry and develop a model that could be used by both small and large aggregate producers to develop a safety management system. This resulted in the publication in December 2005 of the "Core Principles of a Safety Program" by the Alliance. It is available free on the MSHA and NSSGA websites.

At present, the Alliance is working on promoting safety and health through the publication of "rip & share" safety tools in the bimonthly association magazine and articles on timely safety topics for the industry to use in improving their safety programs. MSHA and NSSGA member company representatives jointly develop these tools. The cooperative relationship has made great strides toward improving the safety of the aggregates industry.

You can see this clearly using the data required to be submitted by mine operators on injuries/illnesses and manhours. The attached chart "Comparison of Aggregate Industry Workhours vs. Incident Rates" shows that even with an increasing number of hours worked at aggregates producers' sites there has been a significant reduction in the total incidence rate in the industry. The second chart "Aggregate Industry Incident Rates 1989–2004" shows this data broken down by aggregates industry sector. More progress has been made since 2002 through the cooperative efforts of the NSSGA/MSHA Alliance.

In no way does the NSSGA/MSHA Alliance interfere with the compliance program of the agency. MSHA has an important role in ensuring that safety at aggregates mines and quarries maintain standards that protect employees. The MSHA enforcement program operates independently of any of the cooperative industry alliances. The Mine Safety Act, unlike any other safety agency, requires complete inspections of every mine property 2 or 4 times per year depending on whether it is surface or underground, respectively.

The mining industry is more heavily regulated and inspected than general industry covered under the Occupational Safety and Health Administration regulations. It is important that caution be exercised before rushing to impose more regulations on the mining sector. Careful study of the programs in place must be made and effective enforcement ensured.

NSSGA believes that the cooperative relationship the aggregates industry has developed with MSHA has led to increased safety for aggregate industry employees. We believe that these relationships rather than being discouraged should be encouraged. They are especially helpful to the small- and medium-sized companies that are unable to afford a staff safety professional by providing the mechanisms necessary for continuous improvement to the safety and health of aggregate workers.

NSSGA appreciates the opportunity to provide comments to this very important hearing.

ADDITIONAL COMMITTEE QUESTIONS

Senator SPECTER. There will be some additional questions which will be submitted for your response in the record.

[The following questions were not asked at the hearing, but were submitted to the witnesses for response subsequent to the hearing:]

QUESTIONS SUBMITTED TO DAVID G. DYE

QUESTIONS SUBMITTED BY SENATOR ARLEN SPECTER

Question. Has the Department made a decision whether to appeal citations invalidated and penalties reduced in the Jim Walters Resources Case?

Answer. On January 17, 2006, Solicitor of Labor Howard Radzely responded to your letter of November 22, 2005, suggesting that the Department should appeal a decision of a Federal Mine Safety and Health Review Commission (FMSHRC) Administrative Law Judge dismissing some of the citations against Jim Walters Resources in connection with the September 2001 explosion at the JWR#5 Mine. The

Solicitor's letter explains that on December 1, 2005, MSHA appealed the decision. A copy of that letter, which explains the specific issues appealed, is attached.

Question. Please provide me documentation of the 16 unwarrantable failure orders the agency issued to Sago Mine in 2005, including specification of what conditions caused MSHA inspectors to issue these orders, and whether any of these conditions that led to the issuance of these orders led to the accident on January 2, 2006.

Answer. The table below offers a description of each of the 16 unwarrantable failure orders issued to Sago Mine in 2005. It is important to remember that the investigation of the Sago Mine accident is ongoing, and the cause of the accident is at this time unknown. In addition, all of these conditions were fully corrected before normal mining activity was allowed to resume in the affected area.

Violation number and Date	Regulation and type action	Condition
7097825—5/20/05	75.400—104 D–1 Citation	Loose coal permitted to accumulate at the No. 3 Belt Drive and Take-up. Condition was obvious and extensive.
7097827—5/20/05	75.360(a)(1)—104 D–1 Order	Inadequate examination of the area cited in the D–1 citation. The mine examiner failed to identify and record this hazard.
7097835—6/02/05	75.340a (1)(i)—104 D–1 Order	Battery charging station was located in the primary escapeway. The air current ventilating the charger was coursed directly to the two working sections instead of to the return entry. Repetitive citation and the conditions were obvious.
7097941—7/12/05	75.203(b)—104 D–2 Order	The mining methods used exposed the miners to hazards related to reduced pillar size. The No. 5 entry was driven off sites to the point that in 60 feet of face advancement the pillar width was reduced to 10 feet. No. 9 entry off sites and cut into the No. 8 entry in only 56 feet of development. Management failed to provide a sight line and exposed the worker to the hazard of overriding forces.
7098153—8/16/05	75.202(a)—104 D–2 Order	Mine roof and ribs were not controlled in the sections where the miners work or travel. There were two prior serious accidents. The conditions were obvious and extensive.
7098154—8/16/05	75.380(d)(1)—104 D–2 Order	Primary intake escapeway from the pitmouth to the No. 3 Belt Head had loose hazardous roof for 50 feet and loose hanging roof through out the area. Also, the travelways were obstructed. This was obvious and extensive and prevented safe passage.
7098158—8/16/05	75.364(b)(1)—104 D–2 Order	Inadequate examination of the primary escapeway for the above order. Management failed to identify and record obvious and extensive hazardous conditions in the weekly examination book.
7098159—8/16/05	75.340(a)(1)(i)—104 D–2 Order	Battery charging station was located in the primary escapeway. The air current ventilating the charger was coursed directly to the two working sections instead of to the return entry. Repetitive citation and the conditions were obvious in an area that required a pre-shift examination to be conducted.
4890534—9/12/05	75.400—104 D–2 Order	Accumulations of loose dry coal were permitted to accumulate around and under the No.6 Belt Drive and Take-up. Extensive and obvious.
4890535—9/12/05	75.400—104 D–2 Order	Accumulations of loose dry coal were permitted to accumulate around and under the No. 5 Belt Drive and Take-up. Extensive obvious, and repetitive.
4890536—9/12/05	74.400—104 D–2 Order	Accumulations of loose dry coal were permitted to accumulate around and under the No. 4 Belt Drive and Take-up. Extensive, obvious, and repetitive.

Violation number and Date	Regulation and type action	Condition
4890537—9/12/05	75.1403—104 D–2 Order	Walkway along the Nos. 6, 5, and 4 Belt Conveyors Belt Drives and Takes-ups were obstructed with deep mud and water. This hazard was obvious, extensive, and repetitive.
4890539—9/12/05	75.360—104 D–2 Order	An inadequate pre-shift examination was conducted as identified in the previous 4 orders. Examiner failed to identify the obvious and extensive hazards and record them in the examination book.
7098364—9/18/05	75.330(b)(1)—104 D–2 Order	Line curtains used to ventilate the Number 7 and 8 entries were rolled, thereby interrupting the face ventilation. No movement of air could be detected with a smoke tube and 0.5 percent methane had accumulated. This condition was obvious. Management was present and allowed this condition to exist.
7149865—11/8/05	75.360(f)—104 D–2 Order	The MSHA inspector issued 6 S&S citations and none of the hazards were identified and recorded in the pre-shift examination book. All of these citations were in the area that was examined by mine management.
7098644—12/14/05	75.400—104 D–2 Order	Obvious and extensive accumulations of loose coal, coal fines, and float dust were permitted to accumulate in the 006 MMU. This was a repetitive condition that was obvious and extensive.

Question. I note that recommendations by MSHA for criminal prosecutions had declined from 38 in 2000 to 12 in 2004; we would like some specification as to what recommendations there [are] on criminal prosecutions.

Answer. MSHA is unaware of the source of information indicating that the agency made 38 criminal referrals to the Department of Justice in 2000, or that the agency made 12 such referrals in 2004. According to the agency's information, the following criminal referrals have been made since 1998, along with the most recent information regarding the disposition of those referrals by the Department of Justice. Please note that the Office of the United States Attorney to whom MSHA makes these referrals may take several years to decide how to proceed on a particular referral.

Please also note that the number of referrals in any given year will vary due to factors beyond the agency's control.

Year	Referrals	Declined	Criminal dispositions [1]	Open cases	Percent declined	Percent criminal dispositions [1]	Percent open
1998	8	4	4		50.0	50.0	
1999	18	13	5		72.2	27.8	
2000	18	8	10		44.4	55.6	
Avg 1998–2000	14.7	8.3	6.3		55.6	44.4	
2001	9	4	5		44.4	55.6	
2002	8	2	4	2	25.0	50.0	25.0
2003	3	1	1	1	33.3	33.3	33.3
2004	3	1	1	1	33.3	33.3	33.3
2005	4		1	3		25.0	75.0
Avg 2001–2005	5.4	1.6	2.4	1.8	27.2	39.4	33.3

[1] Includes pleas, convictions, acquittals, and pre-trial diversions (diverted out of the criminal system).

Question. Is the Acting Assistant Secretary, David Dye, seeking appointment as permanent administrator of MSHA?

Answer. On September 15, 2005, the President nominated Richard E. Stickler of West Virginia to be Assistant Secretary of Labor for Mine Safety and Health.

QUESTIONS SUBMITTED BY SENATOR RICHARD C. SHELBY

Question. As we all know the work of a coal miner is desperately important to the life of our economy. But the job of the coal miner is also extremely dangerous and requires that measures are taken and standards are in place and that those safety standards are enforced. In light of the Sago mine disaster, is MSHA properly staffed and funded to carry out its mission to protect the safety of America's miners?

Answer. Yes. Congress has appropriated sufficient funding for MSHA to carry out its mission. The number of Federal mine inspectors has remained relatively constant over the last ten years, from a low of 902 in 1998 to a high of 986 in 2003.

MSHA ENFORCEMENT PERSONNEL

Year	Coal	Metal and nonmetal	Total
1996	690	289	979
1997	634	272	906
1998	615	287	902
1999	631	318	949
2000	660	305	965
2001	653	326	979
2002	605	339	944
2003	621	365	986
2004	579	371	950
2005	584	357	941

More importantly, the inspection workload has remained steady at 15 mines per inspector in each of the last ten years. The number of coal mines decreased 26 percent over the last ten years but the number of coal mine inspectors declined only 15 percent during that time. Consequently, the coal mine inspector's workload has declined. In the late nineties there were 3.8 coal mines for each inspector. Since 2000, that workload has been reduced to 3.3 coal mines for each inspector.

Although the workload has remained relatively constant, inspection activity, as measured by the average number of hours Federal inspectors spend at each mine, has increased 14 percent in the coal sector to 219 hours per mine over the last five years (2001–2005) compared to 192 hours per mine over the previous five years (1996–2000). MSHA has been able to retain its enforcement capacity through management efficiencies such as optimizing office locations and reducing the paperwork burden on inspectors.

In 2005, MSHA issued the highest number of citations and orders since 1994. In recent years, MSHA increased its use of "withdrawal orders" to gain compliance with the standards. The use of this powerful enforcement tool—withdrawal orders require miners to be removed from the area affected by the order, often disrupting production—increased 17 percent over the last five years when compared to the previous five years. MSHA issued more "withdrawal orders" in both 2004 and 2005 than in any year since 1994.

These increased levels of enforcement have helped reduce mining injury and fatality rates to historic lows. In 1978, the first year the MSHA operated under the new Mine Safety and Health Act of 1977, 242 miners died in mining accidents. Last year, there were 57 mining fatalities, 22 at coal mines and 35 at metal and nonmetal mines. From 2000 to 2005, the mining industry experienced a 33 percent decrease in fatal accidents nationwide—with coal mines seeing a 42 percent decline. The coal mine lost-time injury rate declined one-third over the last five years.

Question. There has been a lot of talk about the communications devices and safety equipment available to mine workers and mine operators. What are the practical applications of these devices and more to the point, do they work? Where are they being used, with success, in deep coal mines?

Answer. Currently, the only MSHA-approved communication systems that are available to mine operators, for use in gassy mines, are mine page phones, leaky feeder communication systems (which use hand-held radios) and the Australian-manufactured PED and Tracker IV TAG Systems. Mine page phones and leaky feeder systems require underground power and interconnecting wires. Communications provided by these systems could be compromised in the event of an emergency, fire or explosion.

MSHA has reviewed several installations of the Mine Site Technologies PED Cap Lamp/Paging system to determine if it is a viable solution to the mining industries' communications concern. There are a limited number of PED installations currently in the United States, and there have been installation and reliability problems with the system reported in the past by mine operators. MSHA investigators have recently visited PED installations at Peabody's Air Quality and Twentymile Mines and Consol Energy's Blacksville and Robinson Run Mines. MSHA has plans to evaluate the PED installation at BHP's San Juan Mine, the only U.S. mine with a surface-installed antenna loop.

MSHA specialists are evaluating the Australian-made Mine Site Technologies Tracker IV TAG system, which is used to identify the general location (within zones) of miners underground. This system is not currently in use in the United States, but is installed in one underground coal mine in Australia. Due to the need for underground power for these two systems, MSHA will evaluate their ability to remain functional after power is interrupted as is often the case after an explosion or mine fire.

One of the significant obstacles that we encounter in dealing with the currently available technology is that most systems are not designed for hazardous location use. Systems intended for use in underground gassy mines, if not properly designed, could themselves present an explosion hazard. The current availability of communications equipment that is designed for hazardous location use is very limited.

In response to MSHA's recent Request for Information (RFI), we have received over 50 proposals from manufacturers for evaluation of emergency communication and tracking systems. Additional proposals are still being received on a daily basis. MSHA is currently reviewing these proposals, arranging meetings with the manufacturers and field testing of the systems. MSHA has posted an internet request for manufacturers of advanced communication and tracking systems to contact MSHA to partner in the development of mine-worthy systems. We have already met with manufacturers of several very promising systems. We are focusing our review of emergency systems on those systems that do not rely on a wire back-bone and that would remain functional in a fire or explosion. Once systems have proven to be mine-worthy and can provide the desired emergency operations, MSHA will assist the manufacturers in obtaining MSHA approval.

Our most current findings on these and other communications and tracking technologies, including the PED and Tracker IV Tag systems, can be found at: http://www.msha.gov/Techsupp/PEDLocatingDevices.asp

MSHA is also working with NIOSH and equipment manufacturers to test and develop thru-the-earth, electromagnetic, locator beacons that may be able to locate miners who have become trapped in the mine. This system may also be able to work with existing proximity systems that MSHA helped develop for underground mines, to prevent miners from being injured by mining equipment. This technology is promising, but needs further development.

MSHA strongly believes that this approach will lead to much improved technology being made available to mine operators to respond to any future emergencies.

Question. What are the protocols or requirements for life support and rescue equipment in a mine?

Answer. At the time of the Sago accident, MSHA regulations required the operator to provide one approved self-contained self-rescuer (SCSR) for each miner who goes underground as well as for each visitor who go underground. An SCSR must have the capacity to protect the individual one hour or longer. On March 9, 2006, MSHA published an emergency temporary standard (ETS) requiring operators to immediately provide an additional SCSR for each miner and to provide caches of SCSRs for each miner at strategic locations in the mine if it takes longer than one hour to exit the mine. Training in evacuation procedures and use of multiple SCSRs is also required. In addition, the ETS requires operators to install lifelines to the surface from working face areas through both escapeways. Finally, the ETS requires mine operators to notify MSHA of mine accidents within 15 minutes.

Question. Are the current laws and regulations sufficient to address the issues, or do we need to take new action?

Anwer. The Mine Safety and Health Act of 1977 is a strong and effective worker protection statute that has contributed to the remarkable reduction in accidents, injuries and illnesses in our nation's mines. The Department has proposed to amend the Act to increase the maximum civil penalty for flagrant violations from the current $60,000 to $220,000 to address egregious violations.

MSHA's comprehensive safety and health standards are constantly being reviewed and adjustments made to improve them or address newly recognized hazards. MSHA issued an Emergency Temporary Standard on March 9, 2006, to address underground supplies of oxygen generating breathing devices, training, lifelines and immediacy of accident notification.

Question. Often we hear of mines being fined and then having those fines reduced. Is this common practice? Does this affect the amount of negligence at the mine?

Answer. If a penalty is contested by the mine operator, it is tried before an administrative law judge of the Federal Mine Safety and Health Review Commission, an independent adjudicatory body. Approximately 6 percent of MSHA's proposed civil penalty assessments are contested each year and 2.5 percent of proposed civil penalties are reduced as a result of this litigation. For the period 1996–2005, the $225 million in proposed civil penalties were reduced by $24 million through litigation. I do not believe the reduction in civil penalties that occurs through litigation materially affects the amount of negligence at mines.

Question. How often does MSHA have enforcement personnel at a facility?

Answer. The Mine Act requires that MSHA make inspections of each underground mine in its entirety at least four times a year and of each surface mine in its entirety at least twice a year. The Mine Act also requires additional spot inspections and investigations to monitor and address conditions such as the liberation of certain quantities of methane, investigations of accidents, and responding to hazardous condition complaints. Therefore, MSHA enforcement personnel are frequently at mines carrying out the inspections and investigations required under the Mine Act.

In 2005, enforcement personnel spent over 685,000 hours at mining operations—about 434,000 hours at coal mines and 251,000 hours at metal and nonmetal mines. There were MSHA enforcement personnel at mine sites every day last year.

The amount of time enforcement personnel spend at any particular mining operation is contingent on a host of factors, including the size and type of the operation, the condition of the mine, and the number of days the mine is open or in production. Because underground coal mines are generally more hazardous work environments than other types of mines, MSHA enforcement personnel spend more time at underground coal mines than at other types of operations.

In 2005, MSHA spent, on average, over 60 days at each underground mine that produced coal during at least part of the year. On many of these days there were multiple inspectors at these mines. There were over 100 underground coal mines at which MSHA enforcement personnel spent over 100 days on site, with 16 of these mines having MSHA personnel on site more than 200 days during the year.

Question. Explain the process for enforcing regulations and any subsequent appeals process for fines levied.

Answer. MSHA conducts frequent inspections of the Nation's mines to ensure compliance with MSHA's safety and health standards. If an inspector finds a violation of a standard, he must issue a citation or, in certain specified circumstances, a withdrawal order. In addition, MSHA, applying statutory criteria, proposes a civil penalty for every violation. If a citation, order, or penalty is contested by the mine operator, it is tried before an administrative law judge of the Federal Mine Safety and Health Review Commission, an independent adjudicatory body. The judge's decision may be appealed to the 5-member Commission, and the Commission's decision may be appealed to a U.S. Court of Appeals. Although the judge and the Commission are required to apply the same penalty criteria as MSHA, they are not bound by MSHA's penalty proposal. In contested cases they frequently reduce MSHA's penalty proposal, often significantly.

Question. What is MSHA doing to ensure that we continue to reduce the risk of mining to miners?

Answer. MSHA is relying on all its statutory tools to ensure that we continue to make mining a safer occupation. MSHA is committed to strong, fair and consistent enforcement. For example, last year MSHA completed 99.6 percent of its statutory coal mine inspections and continued to respond to every hazard complaint received. MSHA also has an active training and education program including an office devoted to assisting small mines with their compliance responsibilities. In addition, compliance assistance is available to all mine operators who face challenges in understanding mining's risks and dealing with compliance issues across the board. MSHA is also supporting research efforts by NIOSH into new technologies for improved mine safety and health. In addition, MSHA is using its own resources to test and develop new technologies for mine emergencies including use of robots and sensing devices that will de-energize equipment before it can injure or kill a miner working in close proximity.

QUESTIONS SUBMITTED BY SENATOR ROBERT C. BYRD

Question. Why did it take two hours for MSHA to be notified of the Sago explosion?

Answer. The mine operator is responsible for immediately notifying the appropriate MSHA district office once an accident occurs, as required by MSHA regulations at 30 CFR 50.10. It should be noted that MSHA provides a 24/7 emergency answering service number for the operator to use if the operator is unable to reach appropriate district personnel. The emergency number is listed in 30 CFR 50.10 and on MSHA's website and is regularly made known to mine operators through various methods. When the mine operator calls the emergency toll-free number, the answering service immediately contacts the appropriate MSHA headquarters official who then immediately notifies the district office. This service has been provided to the mining industry for many years.

The management at the Sago mine had the telephone numbers of persons in the district that had jurisdiction over the mine. According to MSHA's telephone logs, the mine operator began to call MSHA personnel at approximately 8:15 a.m. on January 2, 2006. At approximately 8:30 a.m. the mine operator was able to speak with an MSHA field office supervisor. Sago management also did not contact MSHA's emergency answering service number. The issue of immediate notification is currently being examined by MSHA's accident investigation team and enforcement action will be taken if appropriate. On March 9, 2006, MSHA published an emergency temporary standard that, among other things, defined the requirement to immediately notify MSHA of accidents by clearly stating that such notification must be without delay and within 15 minutes.

Question. Given the 2.5 hours it took MSHA to be notified of the underground Alma fire, what is wrong with MSHA's notification system?

Answer. Under MSHA regulations, the mine operator is required to immediately notify the Agency when an accident occurs. If the operator fails to immediately contact MSHA, the Agency's response and rescue efforts suffer accordingly. The management at the Alma mine had the telephone numbers of persons in the district that had jurisdiction over the mine. As noted above, Alma mine management also had access to the Agency's 24/7 emergency answering service number. MSHA's accident investigation team has begun a careful and thorough process to examine this issue and will issue a report of its findings and enforcement action will be taken if appropriate. On March 9, 2006, MSHA published an emergency temporary standard that defined the term "immediate" to mean "at once without delay and within 15 minutes." In the event that communications are lost due to an emergency or other unexpected event, the term "immediate" means "at once without delay and within 15 minutes of having access to a telephone or other means of communication."

Question. Why did it take six hours for rescue teams to arrive at Sago?

Answer. It was the responsibility of Sago mine management to provide the services of mine rescue teams for the rescue and recovery effort. MSHA regulations at 30 CFR 49.2(f) state that the mine rescue station(s) with which the two teams required to be available at each mine are associated must be located within two hours ground travel time from the mine. In addition, 30 CFR 49.2(g) requires that the teams present themselves at the mine within a reasonable time once notification is given. The explosion occurred at approximately 6:30 a.m.; MSHA was notified at approximately 8:30 a.m.; and the first two mine rescue teams arrived at approximately 11:00 a.m., or 4.5 hours after the explosion. The teams were prepared to go underground at noon. However, even with mine rescue teams on the premises, the conditions within the mine had to be determined safe enough for their entrance (e.g., because of possible elevated carbon monoxide levels and the risk of secondary explosions due to high methane levels). At approximately 5:20 p.m., teams were sent into the mine to begin rescue work after MSHA and State officials determined that this work could safely commence. The time that the two mine rescue teams were notified will be determined during the accident investigation. Each distinct response issue is currently being investigated by MSHA's accident investigation team, which will produce an independent report of its findings, and enforcement action will be taken if appropriate.

Question. Why is a rapid notification and response system not available?

Answer. Under MSHA regulations, the mine operator is required to immediately notify the Agency when an accident occurs. If the operator fails to immediately contact MSHA, the Agency's response and rescue efforts suffer accordingly. The management at the Sago mine had the telephone numbers of persons in the district that had jurisdiction over the mine. As noted above, Sago mine management also had access to the agency's 24/7 emergency answering service number. This emergency number has been published in 30 CFR Part 50 for many years.

A mine rescue team's availability in relation to mine emergency response is governed by MSHA's regulatory requirements. Coupling a reasonable period of time to get to the mine rescue station with a 2-hour ground travel distance from the mine rescue station, one could realistically expect a 2- to 3-hour response time by mine rescue teams. Underground fires and explosions result in conditions that must stabilize over a period of time before rescuers can safely enter the mine. Under MSHA's mine emergency response requirements, mine rescue teams almost always arrive at emergency sites before the conditions are safe enough for them to enter the mines. MSHA's recent Request for Information (RFI) asks for comments from the public on mine rescue procedures and mine technologies.

Question. Why does MSHA not require more than two rescue teams at every mine in case of emergency?

Answer. The legislative history for section 115(e) of the Mine Act indicates that Congress considered the ready availability of mine rescue capability in the event of an accident to be a vital protection for miners. In responding to the direction of Congress in section 115(e) of the Act, the rulemaking process addressed, and the final Part 49 rule required the ready availability of two mine rescue teams for each underground mine. In the discussion and summary of the final rule for Part 49, MSHA addressed hundreds of comments and statements.

Since its inception, compliance with Part 49 has been accomplished by company teams, contract teams and state-maintained teams. Historically, the industry's response to accidents has included sending additional mine rescue teams, equipment, and personnel to emergency sites. This additional response is voluntary and not mandated. Notably, there was no shortage of well-trained mine rescue teams at the Sago and Alma mines. Following the Sago and Alma tragedies, MSHA is evaluating the current needs and response capabilities for effective mine rescue.

It is important to remember that mine rescue teams using breathing apparatus perform a very specialized function and are only used in conditions where there is irrespirable air as the result of a fire or explosion. The vast majority of mining accidents are handled without the need for mine rescue teams.

Question. Given how many rescue teams were borrowed at Sago, how prepared is a coal operator who meets MSHA's minimum requirement of two rescue teams per mine?

Answer. Two mine rescue teams are able to conduct the initial phases of a rescue and recovery operation. One team conducts mine exploration while the other team backs them up at the fresh air base. If the emergency event extends beyond a period of 8 hours additional teams are essential. Historically, most mine emergency events last well beyond 8 hours. The industry presently relies on the rest of the mining community to provide additional teams necessary to respond to emergency events. This has proven to be a reliable process in the past.

Question. What would have happened if other operators were not able to loan rescue teams?

Answer. Historically, the mining industry has demonstrated their willingness to assist other companies in an emergency situation. Inability or refusal to lend rescue teams to other mines has simply never occurred. If other operators had not responded to the Sago disaster, the company would have had to rely on State teams in West Virginia and other state maintained teams. In addition, MSHA's Mine Emergency Unit (MEU) would assist these teams in the rescue operations.

Question. Given the number of rescue teams used at the Sago and Alma Mines, how stressed is the system right now? What happens if another mine rescue should be necessary in the coming days?

Answer. An adequate number of teams responded to the Sago and Alma accidents. If another accident were to occur that required mine rescue teams to intervene, history has shown the teams that responded to Sago, Alma, and/or other teams would be able and willing to respond. There are almost 190 mine rescue teams throughout the United States, available to respond to mine emergencies when needed.

Question. In 2002, why did MSHA acknowledge the need to increase the number of mine rescue teams, but then withdraw its assessment of current regulations?

Answer. The number of mine rescue teams has declined over the years, as has the number of mines. MSHA looked at regulations that would increase the number of these teams and held a public meeting in March 2002 in Barbourville, Kentucky to gather current ideas and suggestions concerning mine rescue capabilities and preparedness. Both labor and industry stated that cost is the major factor considered in establishing a mine rescue team. Recommendations to MSHA focused on incentives, particularly reducing penalties for violations if a mine had a mine rescue team. Legally, MSHA could not adopt that approach. Therefore, MSHA withdrew the mine rescue agenda item and issued two Program Information Bulletins that addressed mine rescue cost concerns related to training and technical assistance.

Question. Without regulations, how can MSHA address recalcitrant coal operators who comply only with the minimum requirements for mine rescue?

Answer. Mine operators who comply with the legal requirements of the Mine Act and MSHA regulations are meeting the requirements of the law and are providing the degree of protection required by the Mine Act. MSHA is actively developing regulations to ensure that miners are protected during a mine emergency. For example, on March 9, 2006, MSHA published an emergency temporary standard requiring operators to immediately provide an additional SCSR for each miner and to provide caches of SCSRs for each miner at strategic locations in the mine if it takes longer than one hour to exit the mine. Training in evacuation procedures and use of multiple SCSRs is also required. In addition, the ETS requires operators to install life-lines to the surface from working face areas through both escapeways. Finally, the ETS requires mine operators to notify MSHA of mine accidents within 15 minutes.

In addition, on January 25, 2006, MSHA published a Request for Information (RFI) on issues related to mine rescue teams and mine rescue equipment and technology. MSHA is in the process of evaluating the comments in response to the RFI and has in-mine testing underway for several promising communication systems to aid in mine rescue.

Question. Why did MSHA allow the Sago Mine to continue to operate with a total of 276 safety violations?

Answer. MSHA does not have the authority to force a mine to permanently cease operations. Under certain circumstances, MSHA has the authority to suspend mining activity in the area affected by a violation by issuing a "withdrawal order." MSHA aggressively exercised its withdrawal authority against Sago Mine in 2005, issuing 18 separate "withdrawal orders" that shut down mining activity in specific areas of the mine until health and safety problems were corrected by the operator. Under the Mine Act, the operator of the Sago Mine has the right to reopen an affected area once all of the safety issues that caused the withdrawal order have been corrected. Absent a mine-wide hazard, MSHA does not have the statutory authority to shut down a mine simply because it has a large number of other violations. MSHA records indicate a total of 208 enforcement actions at the Sago Mine in calendar year 2005.

Question. How did MSHA utilize Section 104 of the Mine Act in regard to Sago, which authorizes MSHA to close an area of the mine, after 90 days notice to a coal operator, if MSHA finds an unwarrantable failure to curb habitual violations?

Answer. MSHA utilized section 104(a) citations, section 104(b) withdrawal orders, section 104(d) citations, section 104(d) unwarrantable failure withdrawal orders, and section 103(k) withdrawal orders at the Sago Mine. MSHA did not use its "pattern of violations" withdrawal order authority at the Sago Mine because the mine did not meet the criteria for a "pattern of violations" under the Agency's pattern regulation at 30 CFR Part 104. This regulation requires a review of the mine's compliance records at least once a year for: (1) a history of significant and substantial (S&S) violations; (2) use of increasingly stringent enforcement tools; and (3) evidence of the operator's lack of good faith in correcting the problem that results in repeated S&S violations. In the preamble to the pattern regulation, MSHA suggested that a prior two-year history be used in conducting the pattern review. The Sago mine did not operate continuously during 2004, making a review for the period of 2004 to 2005 problematic. In addition, the Sago mine changed ownership during 2005, with changes in company personnel. The accident incidence rate at the mine declined late in 2005, under the new owner. The fact that Sago received 208 enforcement actions during 2005 would not, alone, trigger the pattern sanction. The compliance record had to be reviewed within the context of other pattern criteria. MSHA had appropriately increased enforcement at the Sago mine using the tools available under sections 103, 104, and 107 of the Mine Act.

Question. How many pattern letters or notices did MSHA send in regard to the Sago Mine?

Answer. None. The mine did not meet the criteria for a "pattern of violations" specified in 30 CFR 104.

Question. How many pattern letters have been issued since you became Acting Assistant Secretary in 2004?

Answer. Since I became Acting Assistant Secretary, MSHA has issued one pattern letter. Subsequently, the mine operator submitted an action plan to improve safety at their mine. Evidence showed that improvements did occur and that the operator was committed to improving safety and health conditions.

Question. How did MSHA utilize Section 108 of the Mine Act in regard to Sago, which authorizes MSHA to institute a temporary injunction, whenever MSHA finds a mine operator engaged in a pattern of violations?

Answer. No such injunction has ever been sought or issued in the history of the Mine Act. In any case, the mine did not meet the criteria for a "pattern of violations" specified in 30 CFR 104.

Question. How many times has this authority been invoked since you became Acting Assistant Secretary in 2004?

Answer. No such injunction has ever been sought or issued in the history of the Mine Act.

Question. What influence does the Arlington office have on the size of penalties assessment by inspectors? Since 2001, have policies been enacted that allow the Arlington office to veto or reduce the size of penalties? Has the Arlington office ever vetoed or reduced the size of a penalty assessed by an inspector?

Answer. MSHA's inspectors do not determine the amount of civil penalties for the citations and orders they issue. Almost 98 percent of all civil penalties issued are assessed by MSHA's Office of Assessments using its computerized assessment system based on criteria in 30 CFR Part 100. The inspectors establish the specific values for each of the five Part 100 criteria when they issue citations and orders. Thereafter, the citation and order data is entered into MSHA's electronic data system and subsequently transferred to Assessments for civil penalty determinations. The remaining 2 percent to 3 percent of civil penalties are for the more egregious violations that receive special assessments. These assessments are higher fines determined by special assessors in the Office of Assessments using procedures to evaluate the same five criteria given in 30 CFR Part 100.

MSHA has not adopted any policies since 2001 that allow the Arlington office to veto or reduce the size of penalties. All penalties are determined by the Office of Assessments. The Office of Assessments does not seek a consensus among senior agency officials when determining penalties. I have never reduced an assessment proposed by the Office of Assessments.

When a civil penalty is issued, the mine operator has 30 days to pay or contest the assessment. Once a proposed penalty is contested, the Mine Act mandates that the proposed assessment shall not be compromised, mitigated, or settled except with the approval of the Mine Safety and Health Review Commission (Commission). In addition, no penalty assessment, which has become a final order of the Commission shall be compromised, mitigated, or settled except with the approval of the Court. Therefore, reductions of civil penalties are made by or approved by the Commission and its administrative law judges.

Question. MSHA announced on Saturday that it would reassess regulations related to mine rescue teams and technology. Does this announcement mean that MSHA made a mistake in abandoning its assessment of these rules in 2002?

Answer. MSHA recently published a Request for Information on Mine Rescue Equipment and Technology. We did so because significant questions remain about the availability and reliability of current technology. Promising technologies are known to us, but practical experience with their use remains very limited. Our decision to explore these technologies and their potential application in the nation's mines is not inconsistent with the decision to remove the previous rulemaking agenda item, which was never published as a proposed rule. Recommendations to MSHA focused on incentives, particularly reducing penalties for violations if a mine had a mine rescue team. Legally, MSHA could not adopt that approach.

Since 2001 we have had a continuing interest in improving mine rescue capability. We have supported NIOSH research into two-way wireless communications and have worked hard at developing robotic technology for use during underground emergencies. In addition, in March 2002, MSHA held a public meeting at the Mine Health and Safety Academy to gather ideas and suggestions from the mining community on the availability, quality and preparedness of mine rescue teams.

Question. Technology has improved significantly since the days when miners used to carry canaries underground, but mine rescue efforts still are hindered by the inability to communicate directly with miners, to pinpoint their location underground in case of emergency, and to provide them with sufficient oxygen in case they are forced to barricade themselves and wait for rescue.

How can we better coordinate the mine safety technology being developed in other Federal agencies, like the Defense and Energy Departments, and in the commercial world, to equip miners with longer lasting breathing devices and more advanced communications technology?

Answer. In 2005, through two workshops, MSHA and NIOSH reviewed new technology that is currently under development in industry and government, which may have an impact on new Self-Contained Self-Rescuer (SCSR) technology. There are several promising technologies on the horizon including new oxygen sources and filter technology; however, new hybrid SCSR technology may have the most immediate impact on the development of smaller, lighter, longer lasting SCSRs. A hybrid SCSR would combine oxygen generation technology with new filter technology. At least one manufacturer is interested in building such an SCSR. In addition, MSHA and NIOSH would have to update respirator approval and certification standards in order to approve such a novel device. MSHA and NIOSH have found the workshop forum to be useful, and will continue to sponsor such workshops in the future, to further seek out new technology, which is being developed in both government and industry.

MSHA is also working with NIOSH and equipment manufacturers to test and develop through-the-earth, electromagnetic, locator beacons that may be able to locate miners who are involved in emergencies, and who become trapped in the mine. This system may also be able to integrate with existing proximity systems that MSHA helped develop for underground mines to prevent miners from being injured by heavy mining equipment. This technology is promising.

The Mine Act does not provide the Department of Energy (DOE) with statutory authority for research into new technologies for mine safety. However, DOE has partnered with the mining industry to foster the development and use of advanced technologies and processes to save money, cut emissions, and reduce waste. Research and development projects are conducted through this partnership in areas of extraction, beneficiation and processing, materials handling, and cross-cutting applications. MSHA is aware of active and completed projects that have an effect on the mining industry. One project being conducted by DOE, NIOSH, academia and the mining industry is the development of an emergency communication system. This project is one of six selected that MSHA will evaluate from a total of 70 proposals received in response to a request for information recently published. The six selected projects do not rely on a wire back-bone and have the greatest potential to remain functional in the event of a roof-fall, inundation, fire, or explosion.

Security concerns can impair the ability of the departments and agencies within the Department of Defense to share technologies with civilian agencies, particularly in the area of communications.

Question. Why not create an Office of Technology in MSHA to assess new technologies in other Federal agencies and the commercial world, and to see if they would be applicable to mine safety?

Answer. MSHA already performs this function through our Office of Technical Support at MSHA's Approval and Certification Center in Triadelphia, West Virginia. I invite you to visit this facility and see for yourself the high quality of work performed there by MSHA's highly qualified and experienced engineers. Additional work investigating state-of-the-art equipment and technologies related to mine rescue or mine fires and explosions is performed by other MSHA engineers and scientists at the Pittsburgh Safety and Health Technology Center in Bruceton, Pennsylvania. The various technical Divisions within the Centers are constantly looking at new technologies and working with manufacturers, other Federal agencies, academia, and industry to identify, develop, test and promote equipment, systems and other types of products that have the potential to improve miners' safety and health.

Examples of these efforts include a recent partnership with two equipment manufacturers, a mining machine manufacturer, NIOSH, and two mining companies to develop and test a proximity protection system that can be placed on a remote control continuous mining machine to prevent operators from placing themselves in the hazardous area of the machine as it is performing tramming operations. This project was precipitated by the high incidence of pinning and crushing fatal injuries (29) that the Agency has investigated over the last 17 years. The result of this effort will be two commercially available systems that have been approved by MSHA and can be installed on continuous mining machines to prevent these types of accidents from occurring.

Other examples of the new technologies investigated by Technical Support include motion detectors that can be installed in surface mining pits to prevent haul trucks from traveling over berms when dumping, and development of a high impact resistant glass for installation on bulldozers working on surge piles at coal mine preparation plants As coal is removed by draw-off feeders from underneath large coal surgepiles, the coal does not draw to the surface, but rather, bridges over such that the surface appears to be solid. As machinery such as a bulldozer travels over the surface, the bridged-over coal can collapse, and the bulldozer will fall into the void. Fatalities have occurred as coal crushes through the bulldozer cab windows and suffocates the operator. Impact resistant glass will prevent this outcome.

One of the significant obstacles that we often encounter in dealing with the currently available technology is that most systems are not designed for hazardous location use. Systems intended for use in underground gassy mines, if not properly designed, could themselves present an explosion hazard. The current availability of communications equipment that is designed for hazardous location use, for example, is very limited.

In response to MSHA's recent Request for Information on mine rescue technologies, Technical Support has received over 50 proposals from manufacturers for evaluation of emergency communication and tracking systems. Additional proposals are still being received on a daily basis. MSHA is currently reviewing these proposals and arranging meetings with the manufacturers and field testing of the systems. MSHA has posted an internet request for manufacturers of advanced communication and tracking systems to contact MSHA to partner in the development of mine-worthy systems. We have already met with manufacturers of several very promising systems. We are focusing our review of emergency systems on those systems that do not rely on a wire back-bone and that could be destroyed in a fire or explosion. Once systems have proven to be mine-worthy and can provide the desired emergency operations, MSHA will assist the manufacturers in obtaining MSHA approval.

Our most current findings on these and other communications and tracking technologies, including the PED and Tracker IV Tag systems, can be found at: http://www.msha.gov/Techsupp/PEDLocatingDevices.asp

MSHA is also working with NIOSH and equipment manufacturers to test and develop thru-the-earth, electromagnetic, locator beacons that may be able to locate miners who are involved in emergencies, and who become trapped in the mine. This system may also be able to work with existing proximity systems that MSHA helped develop for underground mines to prevent miners from being injured by mining equipment. This technology is promising, but needs further development.

Technical Support scientists and engineers are in constant communication with equipment and product manufacturers and participate in industry consensus standard setting committees to keep abreast of the latest state-of-the-art technologies that may offer improvements in miner's safety and health. We strongly believe that this approach leads to much improved technology being made available to mine operators to improve mine safety and health.

Question. The 1977 Mine Act Committee report reads: "To be successful in the objective of including effective and meaningful compliance, a penalty should be of an amount which is sufficient to make it more economical for an operator to comply with the Act's requirement than it is to pay the penalties assessed and continue to operate while not in compliance." How did the severity of the penalties change as the number of citations issued against the Sago Mine began to increase?

Answer. From the beginning of calendar year 2004 through the end of calendar year 2005, the number of citations and orders issued to the Sago mine increased. More importantly, the severity of the enforcement actions issued to the mine increased significantly later in calendar year 2005. During the last half of 2005 thirteen section 104(d)(2) withdrawal orders were issued to the Sago Mine. Only three section 104(d) citations and orders had been issued to the mine during the previous 18 months. As a result, the amount of civil penalties also increased for the mine from $9,515 during CY 2004 to over $20,300 during the first three quarters of CY 2005 and the majority of the unwarrantable failure withdrawal orders have not yet been assessed a civil penalty.

Question. Given the increase in coal prices and profits, is it more economical for coal operators to comply with the Mine Act or to pay a penalty?

Answer. The Administration believes that the statutory cap on civil penalties is too low to deter repeat and flagrant violations of the Federal Mine Safety and Health Act, and has urged the Congress to increase the statutory cap from $60,000 to $220,000 for flagrant violations. MSHA has also announced its intention to revise the decades-old civil penalty system.

Question. If penalties are large enough, why did the number of citations at Sago Mine continue to increase?

Answer. MSHA inspectors are instructed to cite all violations of the Federal Mine Health and Safety Act, and, in the last half of 2005, 13 unwarrantable failure withdrawal orders were issued. Compliance at the Sago Mine was improving in the last quarter of 2005, but we were still not satisfied with their safety performance and were devoting considerable MSHA management time to addressing the problem with company officials.

Question. The Mine Act provides mine operators with the right to contest a fine, and penalties often are negotiated. How often are penalties negotiated?

Answer. Civil penalties are proposed by MSHA and are not negotiable. The citations and orders can be contested to, and the penalties reduced by, the independent Federal Mine Safety and Health Review Commission. For the period 1996–2005, roughly 6 percent of all proposed assessments issued by MSHA were contested, and the Commission reduced 2.5 percent of all proposed assessments.

Question. What is the difference between the fines assessed and the amounts collected?

Answer. For the period 1996–2005, approximately $225 million in proposed civil penalties were assessed to mine operators and contractors nationwide, that number was reduced by $24 million through litigation at the Federal Mine Safety and Health Review Commission, and MSHA has received payments totaling over $158 million, which is an 80 percent collection rate. When a civil penalty is contested and subsequently reduced through litigation, the collection rate is over 98 percent.

Question. To what extent are the negotiations a matter of public record and to what extent do off-the-record negotiations take place?

Answer. All settlements in contested civil penalties must be approved by the Federal Mine Safety and Health Review Commission and are therefore a matter of public record. All records related to the rare cases when the Agency reduced an assessed civil penalty due to adverse effect on an operator's ability to stay in business are available to the public in MSHA's case files.

Question. What was the chain of command at the Sago and Alma Mines?

Answer. The chain of command at any disaster, including Sago and Alma, changes as senior officials arrive on property. Subsequent to the initial arrival of MSHA first response personnel, the District Managers arrived on the properties and assumed their roles as senior MSHA officials for the emergency operations. The District Managers remained in charge of MSHA's efforts until recovery operations were completed.

The Administrator for Coal was at the Sago and Alma mine sites during the emergency operations. He provided guidance and assistance to the District Managers during the mine emergency operations.

The Acting Deputy Assistant Secretary arrived at the Sago property on the second day and was the senior MSHA official on property. He attended briefings for families, met privately with families, and provided briefings for the media and headquarters personnel.

During both incidents, I was located at MSHA's headquarters command center where I received hourly updates from on site personnel and participated in decisions as needed and provided coordination with senior DOL officials.

Question. How were decisions made?

Answer. A three-party group was established at each of the operations. The group consisted of designated personnel from the Company, State, and MSHA. As issues arose, the group discussed options and came to consensus on decisions. A senior mine official, MSHA's District Manager, and the head of West Virginia's mine safety department had oversight responsibilities for this group to ensure that appropriate, timely, and safe decisions were made. As appropriate, MSHA's District Manager consulted with other senior officials prior to presenting an option to the three-party group.

Question. Who had the final authority to make decisions?

Answer. The Mine Operator was responsible for making decisions concerning the mine and the emergency operations. MSHA issued a section 103(k) order for the mine and the State of West Virginia issued a similar order, requiring the operator to submit a plan to MSHA and the State for recovery. The plan had to be approved by MSHA and the State before implementation. By its very nature, this process created a problem-solving environment. Decisions at the two mine emergency operations were made through discussion and consensus among the participating parties, with each bringing their experience, knowledge and skills to the process.

Question. What do we know about the MSHA rule issued in 2004 regarding belt entries and the fire at the Alma mine?

Answer. We know that the last approved map for the Alma mine indicated that belt air was being used to ventilate the longwall section, but not the section where the two deceased miners worked. In May 2000, the mine operator submitted a petition for modification seeking permission to use belt air to ventilate workings. The petition was granted in June 2000, with stipulations. Most of these stipulations were included in the 2004 belt air rule, which then rendered the petitions granted prior to the effective date of the rule unnecessary.

While it is generally known that smoke from the fire contaminated the escapeway being used by the crew from number 2 Section at Alma Mine, MSHA is not able at this time to confirm the reasons for this contamination because the matter remains under investigation, and all facts are not yet known. There is no evidence that the proper use of belt air, as defined in the Alma mine's approved ventilation plan, in any way hindered the escape of the miners.

Question. How many mines use belt entries to ventilate workplaces?

Answer. Currently, there are 41 coal mines in the United States that use air from a belt air course to ventilate a working section. MSHA began granting petitions for modification to permit the use of belt air for this purpose in 1980. Sixty-seven petitions were approved between 1993 and 2001, each with a specific finding that the practice was safe. An additional 27 were approved between 2001 and the publication of the Belt Air rule in 2004, which included the major stipulations of previously granted petitions and rendered those petitions invalid with the application of a nationwide rule.

MSHA's 2004 belt air rule not only codified what had become a routine and safe practice with the safeguards of technologically advanced atmospheric monitoring systems (AMS) and multiple fire suppression systems, but increased miner protection by adding various requirements that were not included in all the petitions. For example, all sensors used must be listed by a Nationally Recognized Testing Laboratory; the trunk lines for the communication system and the AMS must be installed in separate entries; carbon monoxide sensors must be installed in the intake escapeways; sensor spacing must be reduced to 1,000 feet (as opposed to 2,000 feet for mines using older petition requirements); and lifelines are required when returns are used as alternate escapeways. MSHA records indicate that there has never been a fatality caused by belt air carrying fire or contaminants to the working section under the current rule or any of the preceding petitions for modification.

QUESTIONS SUBMITTED TO BENNETT K. HATFIELD

QUESTIONS SUBMITTED BY SENATOR ROBERT C. BYRD

Question. Why did it take two hours to notify MSHA of the explosion at Sago? Is MSHA's notification system adequate?

Answer. My written statement presented to the Subcommittee on January 20 provides significant detail on this point. However, to address the question more concisely, the time lapse of approximately 117 minutes consists of a 69 minute period (6:31 to 7:40 A.M.) when the notification process wasn't triggered because mine management was fully engaged in the initial assessment and emergency response, and a 48 minute period (7:40 to 8:28 A.M.) during which notification was being actively attempted but many telephone calls were not being answered.

The initial 69 minute period reflects the time during which Sago mine management was: (i) attempting to determine what had happened, (ii) safely evacuating the 15 miners from underground that were not trapped, and (iii) launching a spontaneous rescue effort to reach the remaining 13 miners. The 5-man team of mine supervisors and safety managers reached over two miles into the mine, reestablishing temporary ventilation as they advanced to replace the explosion-damaged infrastructure, before being turned back by a wall of thick black smoke and heightened explosion risk when they were only 1,000 to 1,900 feet from the trapped crew.

The subsequent 48 minute period reflects the time required to successfully make direct contact with a Federal (MSHA) inspector—although it should be noted that a state inspector was contacted within 16 minutes. Multiple calls were placed during this period but state and Federal offices were closed and many people were traveling due to the holiday.

—The first notification call went to the state office in Fairmont at approximately 7:40 A.M. Subsequent calls to home telephones and cell phones of state mine inspectors were finally successful with a returned call from state inspector John Collins at 7:56 A.M.

—The notification calls to MSHA contact numbers began at approximately 7:50 A.M. with a call to the home of MSHA Field Office Supervisor Kenny Tenney. This call went to an answering machine. Subsequent calls to MSHA's District 3 office, other inspector home telephones, and inspector cell telephones were also unanswered. At approximately 8:28 A.M., MSHA inspector Jim Satterfield was contacted via his home telephone and was notified of the accident.

Hence, the notification process for state mine regulators required approximately 16 minutes, for a total elapsed time from accident occurrence of 85 minutes. The notification process for MSHA required approximately 48 minutes, for a total elapsed time from accident occurrence of 117 minutes.

With respect to the question as to whether MSHA's notification system was adequate, we note that January 2 was a State and Federal holiday so it was admittedly more difficult to reach officials than on a normal workday. A central emergency call center would have made it easier for the company to "notify" MSHA and the state. However, a central call center would probably have had the same difficulty we experienced in reaching individual officials outside of normal working hours. Thus, even with a fully staffed central call center, we cannot say that mobilization of state and Federal officials would have resulted in a quicker response. Likewise, earlier contact with those officials would almost certainly not have hastened entry of the rescue teams into the mine due to the trend of carbon monoxide readings confirming high risk of a second explosion.

Put simply, based on air quality monitoring by state inspectors beginning at around 8:30 A.M., Federal and State officials were in agreement that it was unsafe to enter the mine. In its deliberation, the Subcommittee should not cast aside the lessons learned by painful experience: rescuers are often at risk from secondary explosions and additional lives should not be put at significant risk in a rescue effort.

Question. Why did the Sago Mine have only two rescue teams available in case of emergency? What would have happened if other coal operators were unable to loan rescue teams?

Answer. To begin with, it is important to recognize that the availability of mine rescue teams had no bearing on the tragedy that occurred at the Sago Mine. The mine rescue teams responding on January 2 had to spend hours and hours waiting as State and Federal mine authorities monitored carbon monoxide levels to determine whether safe entry was possible.

That said, each mine is required by applicable law to have emergency coverage by two mine rescue teams within two hours of the mine. The Sago Mine contracted with the Barbour County Mine Rescue Team to provide this service. These teams included several miners from Sago and its affiliated operations.

To answer the hypothetical question about what would have happened if other coal operators had been unable to loan rescue teams, there are many mine rescue teams that are not associated with any particular coal company—including the Barbour County Mine Rescue and Tri-state Mine Rescue groups—both of which assisted in the Sago Mine accident. More importantly, we are aware of no instance where there has been any widespread refusal or inability of other coal operators to offer the services of their mine rescue teams. Indeed, the events of the past month have proven that, even in an environment of stiff competition in the coal industry, competitors quickly rush forward with their mine rescue teams to assist in the time of a mine emergency. For example, in the Sago Mine accident, a total of 15 mine rescue teams were assembled on site. Additional to the Barbour County and Tri-state teams referenced earlier, this group included several mine rescue teams from Cansol Energy, one of our primary competitors in the region, as well as ICG's own Viper Mine Rescue Team that flew in from Illinois. Indeed, we received more offers of mine rescue assistance than we were able to utilize—including those from two other ICG competitors—Massey Energy and Arch Coal.

These demonstrations of volunteerism, along with other recent examples across the coal industry, clearly confirm that overwhelming response by the industry in the time of need is the norm and not the exception.

Question. How large are the fines assessed by MSHA in comparison with your companies [sic] profits? What is more economical—to comply with the Mine Act, or to pay the fine and operate in violation?

Answer. During the first nine months of 2005, ICG (including the earnings of subsidiaries it acquired from Anker Coal Group in November 2005) had operating profit (EBITDA) of approximately $81 million. During that same period, the fines assessed by MSHA against the associated mining operations were approximately $116,000. However, it is important to note that the Saga Mine lost approximately $6.2 million during that same nine month period. Despite the poor financial performance, upon ICG's gaining management oversight, our company initiated approximately $5.0 million in capital expenditures to improve the Sago Mine. Many of those capital expenditures had a direct impact on safety. Those safety efforts were detailed in my written and oral testimony to the Subcommittee.

With respect to the nexus your question places between profits and the size of fines, I wish to point out that profitable mines would more logically have the means to maintain the capital expenditures that would allow more safety enhancements. However, some unprofitable mines—like Sago—are subsidized by their owners to make such improvements. The Sago Mine currently operates solely to meet the contractual requirements of its customers. We certainly hope that the Sago Mine becomes profitable, but the facts clearly reflect that while it has operated (even at a loss), ICG has provided an influx of capital. Comparing fines with profits seems to draw the incorrect correlation—fines should be based upon the severity of the violation rather than the financial health of the company cited.

The second question also reflects an incorrect premise. For ICG and its subsidiaries, the question is not whether mining can be conducted more economically by operating in violation of the law. ICG and its subsidiaries are committed to operating in compliance with the Mine Act. Unlike a factory floor, an underground mine is a dynamic environment, subject to ever-changing geologic conditions, in which a combustible fuel—coal—is mined. We proactively correct conditions without governmental oversight, as is shown by the steps we have taken as discussed in my testimony, but we also correct cited conditions immediately upon receiving a citation.

Despite the apparently incorrect premise of the second question, safety and productivity go hand in hand. If one ignores the moral, social, and legal obligations that are the principal controlling factors for our company, the true economic disincentives to operating a non-compliant coal mine include: (i) Disruptions to production caused by governmental orders to cease production, even if limited in duration, and the loss of revenues due to that disruption; (ii) Expense related to lawsuits; (iii) Increased workers compensation and insurance premiums; (iv) Damage to reputation among customers and competitors; and (v) Inability to attract and retain skilled workers that clearly have other employment alternatives in the current labor market -all in addition to fines and penalties. Of course, the ultimate disincentive is the risk of injury or death to our employees.

The true economic decision is not whether to operate a mine unsafely; rather it is whether a mining project is economically feasible. No mine operator will choose to operate at a loss indefinitely—nor will it choose to invest in new mines that are not economically justifiable. However, speaking for ICG, compliance with all applicable safety standards is the fundamental principle of our business; investments have been and will be analyzed only in a manner that presupposes such compliance.

QUESTION SUBMITTED TO CECIL E. ROBERTS

QUESTIONS SUBMITTED BY SENATOR ROBERT C. BYRD

Question. At MSHA's Emergency Preparedness Conference in 1995, industry and labor representatives voiced concerns about the dwindling number of rescue teams allotted per mine.

How many mine rescue teams should be maintained at each mine?

Answer. The Union believes that every operator of a coal mine should be required to maintain two qualified mine rescue teams, which would be immediately available whenever miners are in the underground workings of the mine. (In practice, this would probably require the creation of at least three rescue teams so that whenever the members of one rescue team are working their normal assigned job underground, two other teams would be available on the surface, in event of an emergency.) All teams should be comprised of miners who are employed at that mine, as they are acutely aware of the mine's layout and conditions, to enable faster response and to afford their expertise to any additional mine rescue teams that may arrive to assist in an emergency. There should be no exceptions, even for small mines; this requirement must be recognized as a necessary business expense.

In addition, specific training requirements such as drills one weekend each month at the mine site and mandatory participation in mine rescue contests each year should be required.

Question. How do we increase the number of mine rescue teams?

Answer. Allow a transition period of nine months from the date of the legislation, during which time each operator can identify and train sufficient numbers of miners to fill the needed mine rescue teams. Selected miners can be trained through dedicated classes at MSHA's Mine Academy in Beaver, WV and the Lake Lynn Lab in Fairchance, PA. The Rockefeller legislation would provide tax incentives (for a 3-year period) to promote the development of more mine rescue teams.

Question. The 1995 MSHA report on its Mine Emergency Preparedness Conference reads: "the industry's ability to learn from lessons of past accidents is limited by inadequate information sharing as well as by accident reports that do not always reflect what happened."

What else can be done to ensure a complete accident investigation?

Answer. For all multiple-fatalities and life-threatening injuries, and other major disasters like Quecreek, in addition to MSHA's investigation of the accident GAO should review MSHA's conduct.

After MSHA completes every accident investigation (and the GAO completes its report for multiple fatal accidents and other major disasters) the reports should be provided to employees at the District as well as National levels of MSHA, so that employees throughout the Agency can learn from the particular investigation; also copies should be made available to other organizations involved with miners' health and safety, as well as the affected operator and any miners' representatives at the affected operation. These should be generally available to the public, including posting on the MSHA web page.

Also, to ensure all relevant information is relayed to MSHA's accident investigation team, the legally authorized representative(s) of the miners, in accordance with 30 CFR Part 40, should be included in all aspects of the investigation, including all interviews and the physical investigation. As MSHA itself and a Federal judge reviewing the role of miners' representatives at the Sago mine recognized, miners' representatives have the trust of affected miners and act as a valuable conduit of information to those conducting an investigation. The operator should be required to provide information, but not be present when witnesses provide information during MSHA's interviews; excluding the operator will enhance the likelihood that miners will provide complete and accurate information.

Question. How long should it take for MSHA to be notified of an emergency, and how long should it take to organize and deploy mine rescue teams?

Answer. MSHA should be notified "immediately" after a mine operator becomes aware that an emergency condition exists. Immediate should mean no more than fifteen minutes, absent an extraordinary explanation. If in doubt, the operator should provide notice to MSHA even before the scope of the emergency may be fully known. Emergencies for which notice must be provided includes: explosion, fire, innundation, massive roof fall, and a serious injury or fatal accident. The determination to send MSHA field personnel to the site after notification can and must be a determination of the Agency. The mine operator must be responsible for putting its own mine rescue teams into immediate service upon any emergency; it also may seek to call upon additional mine rescue teams from nearby locations even before MSHA arrives on site. However, MSHA must have authority to direct and coordinate the rescue.

The representative of miners' should be notified as soon as notice is provided to MSHA.

By requiring each mine to have two mine rescue teams immediately available, only minimal time would be required to deploy those individuals. Typically a team could be assembled as quickly as 30 minutes, though it could take longer to assemble a team if its members were not all nearby and accessible by phone. Most miners live in the nearby communities and can respond quickly. If all operations maintained at least two teams at all times, then supplemental teams from the nearest mining operations would also be available to assist. Additional rescue teams can arrive at later times to bolster rescue and recovery efforts, and to trade out with the first responders. To give an operator quick access to such additional rescue teams, information about all rescue teams should be easily accessible through MSHA's web page—information about all rescue teams should be regularly updated and MSHA should ensure it is easy to find.

Question. How do we create a rapid response mine rescue system?

Answer. If every operation must maintain at least two mine rescue teams, as we suggest and as is required by the Mine Act, then a "rapid response mine rescue system" is well along to being in place. In addition, MSHA must make its personnel quickly available: phone numbers, including off-hours emergency contact numbers, must be provided to all operators for use in an emergency. All such contact information must be kept up to date. Also, a mobile triage/surgical unit located within a reasonable distance from mining operations could provide valuable resources when rescuers bring survivors to the surface. The Commonwealth of Pennsylvania currently has two such medical/surgical units readily available and can deploy them quickly in the event of a mine emergency. MSHA or the CDC could coordinate central placements of such units.

Question. How do we insure that the size of penalties is large enough to encourage compliance?

Answer. Penalties must be high enough to convince operators that full compliance is the best choice. The current base fine of $60 is far too low. For example, compare MSHA fines with those under OSHA: while OSHA deals with many far less-dangerous industries, fines for knowingly endangering workers begin at $5,000, while a coal operator demonstrating "high negligence" could be fined just a few hundred dollars. The base penalty under MSHA should begin at $1,000, irrespective of the size of the mine, the number of miners exposed or any other factor. For more serious violations, penalties should be much higher. After it assesses a penalty, MSHA should not negotiate or compromise that amount. The Agency must then collect the fines it assesses, on a regular and prompt basis.

There needs to be an expectation throughout the industry that violations of the Mine Act will be cited with regularity, costly for the operator, and that abatement must be accomplished in a timely fashion. Pursuant to the existing penalty structure, the number of miners affected is one of the factors that is considered in assessing penalties. However, even the way MSHA applies this factor is unrealistically low: often MSHA counts how many miners are in the most immediate area without considering that a problem in one area can easily spread harm to all underground miners. Also, MSHA now considers the effect of a fine on an operator's ability to stay in business. We submit that factor is inappropriate. If an operator cannot satisfy its obligations under MSHA, then it should not be in the business. No miner should be subjected to unsafe mining conditions because his employer is on shaky financial footing. Penalties should be the same size for the same violation at all mines.

Fines for a failure to abate should be immediately doubled, then tripled. Repeat violations should generally be assessed much higher amounts. MSHA should also have the authority to shut an operation for noncompliance and for a failure to pay outstanding fines. If an operator continues to ignore the law, it should lose the right to operate.

Question. What do you think of assessing penalties as a percentage of profits rather than a flat rate?

Anwer. This would not be good idea for a number of reasons: many of the smaller and more marginal operations pose some of the greatest risks to miners' health and safety; many coal companies are not publically traded so obtaining information about their profitability may not be available or verifiable; some companies may have artificially low profits because they pay extraordinary compensation to their top executives, or might be motivated to alter their financial records to avoid showing high profits if higher penalties could result; coal companies are notorious for having complicated and quickly-changing corporate structures—figuring out which entity had what profit margin would be difficult to determine and could quickly change.

Question. A miner knows his work area best. The Mine Act guarantees every miner union or nonunion the right to leave the mine whenever he feels his safety is threatened.

How can we ensure that miners feel free to exercise that right without fear of retribution?

Answer. This is a significant issue. In the same way that the NLRA does not prevent employers from firing employees when they try to form unions even though that is illegal, so too, do miners reasonably fear getting fired for exercising their rights under the Mine Act, including the right to withdraw. The right to withdraw is protected through the anti-discrimination provision of the Mine Act, at Section 105(c). However, enforcement of this protection takes too long and is not often pursued.

While MSHA has the authority to seek temporary reinstatement for any miner who is fired unlawfully, this relief is rarely pursued. MSHA should regularly seek temporary reinstatements as soon as it finds prima facie merit to a discrimination claim. Also, it needs to act in a speedy fashion in these cases. MSHA habitually misses its own deadlines in investigating and prosecuting these discrimination cases. It needs adequate funding to accomplish this, and it must make this a high priority. MSHA needs to be more aggressive in protecting all of the miners' rights under the Mine Act.

Note that the right to withdraw has been interpreted to be quite limited: only if a miner believes that his immediate safety is in jeopardy does he have the right to withdraw, and the withdrawal is from the particular area, not the mine itself. To exercise this right, a miner must bring the perceived danger to the attention of his supervisor stating expressly that the miner has immediate fears for his health or safety. The miner can only speak out for his own protection, and cannot seek to have any other miners withdrawn. This right is rarely invoked; to make it more meaningful the legal standard for withdrawal should be liberalized. Finally, a miner who pursues his Mine Act rights is often blackballed, whether or not the case is ultimately successful.

Question. Why are you opposed to the MSHA rule that allows belt entries to be used for ventilation?

Answer. The Union has always been opposed to using belt air to ventilate the working areas. We vigorously opposed implementation of this rule, and sued the Agency in Federal court when it was finalized in 2004; deferring largely to the government (under the applicable "arbitrary and capricious" standard), the court overruled the union's challenges based on the present law.

The belt entry is the only location in the mine where equipment remains in constant motion. This not only includes the conveyor belt, but also head rollers, tail rollers, belt take-ups and structural rollers used to support the conveyor belt. Stuck rollers, coal spillage, worn bearings, unaligned belt conveyor and other problems create extra friction, which can cause fires to ignite on the belts; with coal ever-present, this fuels such a fire. Further, when the belts themselves burn, toxic fumes are also generated.

Conveyor belts extend for miles in underground coal mines, and these areas are not regularly monitored by personnel. The potential for fire in the mine is greatest in the belt entry; MSHA's data demonstrates that mine fires originate in belt entries most often. In addition, the propagation of fire throughout the entry by a burning, moving conveyor belt can quickly make a fire uncontrollable. Air coursing through the belt entry at high velocities further compounds the dangers posed by a belt fire. This is very similar to fighting a forest fire during a period of extremely high winds. The flames are fanned and even a small fire can become uncontrollable quickly.

The end result is obvious. The smoke, fire, and toxic gases are directed by mine design toward the working areas of the mine and directly over miners working at the face. These events can happen quickly, rendering the belt and return entries not viable escapeways. Generally, there will be a third entry that also supplies air to the working section. However, because air pressures in the belt entry are permitted (by MSHA regulation) to be greater than in other entries, the fire and/or fire gases may bleed into other entries adjacent to the belt. Mines with this design allow the other entries to also become compromised when there is a fire on the belt. Miners have difficulty escaping in these cases. Indeed, we understand that t was along a belt entry that was used for ventilation, where the Aracoma Alma No.1 fire began, resulting in the deaths of miners Don Bragg and Ellery Hatfield in January, 2006.

CONCLUSION OF HEARING

Senator SPECTER. Thank you all very much for being here. That concludes our hearing.

[Whereupon, at 1:30 p.m., Monday, January 23, the hearing was concluded, and the subcommittee was recessed, to reconvene subject to the call of the Chair.]

○

APEX
TP_CF
9780160769160